The Greenhouse Delusion
A Critique of "Climate Change 2001"

Vincent Gray

VINCENT GRAY has a Ph.D degree in Physical Chemistry from Cambridge University.

He is a research scientist with a wide experience in five countries (UK, France, Canada, New Zealand and China), in laboratories studying petroleum, plastics, coal, timber, building, and forensic science. He has published widely in the scientific and technical journals, and on Internet websites. For the past 12 years he has specialised in climate science, and is an expert reviewer for the Intergovernmental Panel on Climate Change. He is 80 years old and lives in Wellington, New Zealand.

Vincent Gray, M.A., Ph.D
75 Silverstream Road
Crofton Downs, Wellington 6004
New Zealand
Phone/Fax 064 4 9735939
Email vinmary.gray@paradise.net.nz
©2002

Published by
Multi-Science Publishing Co Ltd
5 Wates Way, Brentwood, Essex CM15 9TB UK
www.multi-science.co.uk

I would like to dedicate this book to John L Daly, whose website "Still Waiting for the Greenhouse" at http://www.john-daly.com, operating from Tasmania, has been the inspiration for genuine scientific discussion on the climate for a whole generation.

Contents

Summary for Policymakers	*1*
1 The History of the Greenhouse Effect	*3*
2 "Climate Change", "Change of Climate" or "Climate Variability"?	*11*
3 Global Warming. What Evidence?	*15*
4 Greenhouse Gases and Aerosols	*35*
5 Sea Level	*51*
6 Computer Climate Models	*57*
7 Forecasting the Future	*67*
8 Extreme Events	*83*
9 Conclusions	*85*
A Note on Sources	*91*

Summary for Policymakers

- Climate has always changed and nothing we can do will stop it from changing.

- There is no credible evidence that the earth is currently warming. Satellite measurements in the lower atmosphere for the past 23 years show no significant temperature change after allowance has been made for natural events. The frequently quoted combined temperature record from weather stations is biased in favour of proximity to cities, airports, buildings, roads and vehicles, all of which have become slightly warmer over the years from increased energy consumption. Surface measurements from remote areas, or from countries with many well controlled sites (such as the USA) show no evidence of significant warming.

- Sea level measurements are even more biased than weather stations. They are mainly near Northern Hemisphere ports, and are subject to local and short and long-term geological changes which are difficult to allow for. Sites in remote, low population places, such as the smaller Pacific islands, show no evidence of recent sea level change.

- The earth's temperature is influenced by greenhouse gases in the atmosphere.

- The changes over the years in the most important of these gases, water vapour, and the clouds that form from it, are virtually unknown.

- The minor greenhouse gas, carbon dioxide, is increasing in concentration, linearly, at the rate of 0.4% a year, and as a result, agricultural and forestry yields are increasing. There are no established harmful effects of the increase in atmospheric carbon dioxide.
- The rate of increase of the only other important greenhouse gas, methane, has fallen steadily for the past 17 years. The actual concentration of methane in the atmosphere is now falling.
- Computer climate models are based on the incorrect belief that changes in greenhouse gas concentrations are the only influences on the climate.
- There are huge uncertainties in the model outputs which are unrecognised and unmeasured. They are so large that adjustment of model parameters can give model results which fit almost any climate, including one with no warming, and one that cools.
- No model has ever successfully predicted any future climate sequence. Despite this, future "projections" for as far ahead as several hundred years have been presented by the IPCC as plausible future trends, based on largely distorted "storylines", combined with untested models.
- The IPCC have provided a wealth of scientific information on the climate, but they have not established a case that increases in carbon dioxide are causing any harmful effects.
- Attempts to suggest a relationship with "unusual" weather events and changes in greenhouse gases have been unsuccessful.

1. The History of The Greenhouse Effect

Jean Baptiste Joseph, Baron de Fourier[1], as early as 1807, suggested that the earth's atmosphere acts like the glass of a hothouse because it lets through the light rays of the sun but retains the dark rays from the ground. The suggestion is included in his *Théorie Analytique de la Chaleur*[2]. John Tyndall in 1858[3] carried out the first reliable experiments on the infrared properties of water vapour and carbon dioxide, resulting in his estimate that water vapour is the key greenhouse gas.

The first extended scientific paper on the greenhouse effect was by Svante Arrhenius[4] in April 1896. Arrhenius was to receive the Nobel Prize for Chemistry in 1903 for his discovery of ionic dissociation.

Arrhenius assumed that water vapour and carbon dioxide in the air are transparent to the sun's rays, but absorb part of the infrared emissions from the earth. He calculated the effect by subtracting the infrared spectrum of the moon, which would undergo absorption, from that of the earth. He then calculated the effect of changing the atmospheric concentration of carbon dioxide. He found that a reduction of one third gave a temperature reduction of just over 3°C. Doubling carbon dioxide gave an increase of 5-6°C Although these results resemble some more recent calculations, the resemblance is fortuitous, as Arrhenius only considered part of the infrared spectrum, and the measurements were crude.

His work was ignored for a number of reasons. The averaged results from weather stations in the Northern Hemisphere (Figure 1.1) showed a fall in temperature for the next 15 years. After that, the world was preoccupied with two world wars and an economic crisis. There were no reliable measurements of concentrations of atmospheric gases thought to be responsible for the greenhouse effect; particularly water vapour and carbon dioxide. This book will show that there are still no reliable measurements of the chief greenhouse gas, water vapour, and that carbon dioxide is not as well characterised as is generally believed.

The work of these pioneers is discussed by Fleming[5] and Baliunas & Soon[6].

In the 1970s a number of scientists put out warnings that a new ice age was imminent. They were based on the supposed mean global temperature as indicated by an averaged record of Northern Hemisphere weather station measurements which showed a decline for 36 years (Figure 1.1, From Stanley[7]).

This record convinced many scientists and journalists that a new ice age was about to commence. The US National Science Board 1972 stated:

"Judging from the record of the past interglacial ages, the present time of high temperatures should be drawing to an end... leading into the next glacial age" (quoted in Matthews[8]).

Rasool and Schneider[9] even argued that greenhouse warming would be overwhelmed by cooling due to aerosols, which "are believed to be sufficient to trigger an ice age."

However, the combined weather station record began to reverse, and move upwards, so a warming crisis was now in order.

On June 23rd 1988 Dr James Hansen, of the Goddard Institute of Space Studies in New York, told the US Senate that he had devised an improved method for calculating the average surface temperature of the earth from the many thousands of temperature measurements collected by weather stations[10].

At about the same time Charles Keeling of the Scripps Institute of Oceanography, University of California, La Jolla, California, provided the first reasonably reliable sequence of measurements of atmospheric carbon dioxide concentration from 1958[11], first from La Jolla, to May

1974, and then from the top of the volcano at Mauna Loa, Hawaii. It showed that the concentration was increasing.

There were developments in the use of computers to model complex systems. An early effort was by Forrester[12] in association with the Club of Rome publication *The Limits to Growth*[13]. Parameters such as population, quality of life, pollution, capital investment and natural resources were parameterized to show that some disastrous future consequences were likely in the near future if current trends continue. The disasters failed to eventuate, however.

Another disaster exercise was the attempt to determine the climatic effects of nuclear war[14]. It concluded that there could be a global "nuclear winter" which could destroy all human and animal life. Subsequent retraction of this conclusion was little publicised, so many still accept the possibility. Fortunately it has not yet been put to the test.

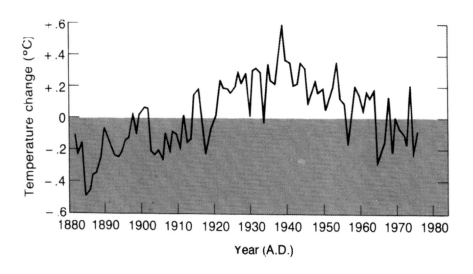

Figure 1.1　　Mean Northern Hemisphere Temperature change from weather station records, as derived in 1975, (from Stanley[7])

Computer climate models based on the theory that all climate change is caused by increases in greenhouse gases then began to appear[15]. They seemed to show that the earth was soon to become uncomfortably warm.

The following diagram (Figure 1.2) shows what was believed to happen[22]. The extra greenhouse gases absorbed additional infrared radiation from the earth, part of which was re-radiated downwards to cause a heating effect.

Because of the public concern, the World Meteorological Organisation and the United Nations Environmental Programme jointly established the Intergovernmental Panel on Climate Change (The IPCC) in 1988. It was set up in order to:

- Assess available scientific information on climate change

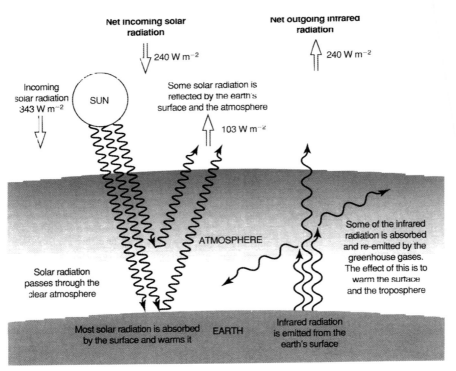

Figure 1.2   The Greenhouse Effect[22]

- Assess the environmental and socio-economic impacts of climate change
- Formulate response strategies

The second and third objectives depend heavily on the conclusions from the first, which, this book will argue, have been wrongly drawn.

Working Group 1, which is responsible for the scientific assessment, has produced three major Reports; 1990[16], 1995[17], and 2001[18] which will provide the main material for this volume.

The Reports consist of reviews of scientific studies on the climate, divided into Chapters. Each chapter has several Lead Authors plus a number of Contributors. The 2001 Report[18] has, in addition, Coordinating Lead Authors, Contributing Authors (and sometimes Key Contributing Authors) and Review Editors. The 1990 Report[16] has an Executive Summary. The 1995[17] and 2001[18] Reports have a Policymakers Summary, and a Technical Summary. The 2001 Report[18] even has, in addition, an Executive Summary for each Chapter. However, none of the Reports has an index, which makes it difficult to find any particular topic, which may often be dealt with in several Chapters.

There were three drafts of each Report, which were circulated to "Expert Reviewers" throughout the world for comment.

This was probably the most extensive scientific exercise ever attempted. The 2001 Report[18] had 15 Review Editors, 124 authors and 397 expert reviewers

The Reports form a valuable compendium of recent work on climate science, but as there is no index to any volume, individual items are often difficult to find. They are often spoiled by unjustified deductions and comments. Comments that were not welcomed by the main authors stood little chance of being considered seriously. The first report has it as follows.

> "Whilst every attempt was made by the Lead Authors to incorporate their comments, in some cases these formed a minority opinion which could not be reconciled with the larger consensus"

The 1990 Report[16] served as the basis for negotiating the United Nations Framework Convention for Climate Change.

Climate Change 01 had two further Reports. The one from Working Group II was called *Climate Change 2001: Impacts, Adaptation and Vulnerability*[19], henceforth referred to in this book as *Climate Change 01: Impacts*. The one from Working Group III was called *Climate Change 2001: Mitigation*[20], henceforth referred to as *Climate Change 01: Mitigation*.

In addition to the three major IPCC scientific Reports there was a 1992 Supplementary Report[21], henceforth referred to as *Climate Change 92*, and "Climate Change 1994: Radiative Forcing of Climate Change and An Evaluation of the IPCC IS92 Emissions Scenarios" henceforth referred to as *Climate Change 94*[22].

References

1. Fourier, J B J. 1822. *Théorie Analytique de Chaleur* (Paris). Translated in 1878 by Alexander Freeman *The Analytical Theory of Heat*. Cambridge University Press.
2. Fourier, J B J. 1827. *Mémoires de l'Académie Royale de Science de l'Institut de France* Tome vii.
3. Tyndall, J. 1865. *Heat a Mode of Motion* 2nd Edn, p 405, London.
4. Arrhenius, S. 1896. "On the Influence of Carbonic Acid in the Air upon the Temperature of the Ground." *Philosophical Magazine* Ser 5, Vol 41, No 251, April 237–276.
5. Fleming, J. R. 1998. *Historical Perspectives on Climate Change*. Oxford University Press.
6. Baliunas, S & W. Soon. 1999. "Pioneers in the Greenhouse Effect." *World Climate Report 4*, (19); http://www.greeningearthsocietyorg/climate
7. Stanley, S. M. 1989. *Earth and Life through Time* page 574, after J.M. Mitchell, in *Energy and Climate; National Academy of Sciences, Washington*. W. H. Freeman & Co New York.
8. Matthews, S. W. 1976. (November) "What's Happening to our Climate?" *National Geographic* 576–615.
9. Rasool, S I & S H Schneider. 1971. "Atmospheric Carbon Dioxide and Aerosols: Effects of Large Increases on Global Climate." *Science*, 173 138–141.
10. Hansen, J & S Lebedeff. 1987. "Global Trends of measured surface air

temperature." *J Geophysical Research* 92 13345–13372.
11. Keeling, C D, R D Bacastow, A F Carter, S C Piper, T P Whorf, M Heimann, W G Mook & H Roeloffzen. 1989. "Analysis of observational data" in *Aspects of climate variability in the Pacific and the Western Americas* (D H Peterson (Ed)) *Geophysical Monograph* 55 American Geophysical Union, Washington DC USA.
12. Forrester, J W. 1971. *World Dynamics*. Wright Allen Press, Cambridge, Mass USA.
13. Meadows, D & M. 1972. *The Limits to Growth*. Universe Books New York, also Pan Books, London.
14. Turco, R P, O B Toon, T P Ackermann, J B Pollack & C Sagan. 1984. "The Climatic Effects of Nuclear War." *Scientific American* 251 (August) 23–41.
15. Schlesinger, M F, & J F B Mitchell. 1987. "Climate Model simulations of the equilibrium climatic response to increased carbon dioxide." *Reviews of Geophysics* 25 760–798.
16. Houghton, J T, G J Jenkins & J J Ephraums (Eds) 1990. *Climate Change : The IPCC Scientific Assessment.* Cambridge University Press. Henceforth referred to as *Climate Change 90.*
17. Houghton, J T, L G Meira Filho, B A Callander, N Harris, A Kattenberg & K Maskell. 1996. Climate Change 1995: *The Science of Climate Change.* Cambridge University Press Henceforth referred to as *Climate Change 95.*
18. Houghton, J T, Y Ding, D J Griggs, M Noguer, P J Van der Linden, X Dai, K Maskell & C A Johnson (Eds). 2001. *Climate Change 2001: The Scientific Basis.* Cambridge University Press. Henceforth referred to as *Climate Change 01.*
19. McCarthy, J J, O F Canziani, N A Leary, D J Dokken & K S White. 2001. *Climate Change 2001: Impacts, Adaptation and Vulnerability.* Cambridge University Press. Henceforth referred to as *Climate Change 01: Impacts.*
20. Metz, B, O Davidson, R Swart & J Pan. 2001. *Climate Change 2001: Mitigation* Cambridge University Press. Henceforth referred to as *Climate Change 01: Mitigation.*
21. Houghton, J T, B A Callander & S K Varney (Eds). 1992. *Climate Change 1992: The Supplementary Report to the IPCC Scientific Assessment.* Cambridge University Press. Henceforth referred to as *Climate Change 92.*

22 Houghton, J T, L G Meira Filho, J Bruce, H Lee, B A Callander, E Haites, N Harris & K Maskell (Eds). 1995. *Climate Change 1994: Radiative Forcing of Climate Change, and An Evaluation of the IPCC IS92 Emission Scenarios.* Cambridge University Press. Henceforth referred to as *Climate Change 94.*

2. "Climate Change", "Change of Climate", or "Climate Variability"?

"Climate" is usually thought of as the average weather, in terms of temperature, precipitation and wind velocity, over a period, and in a particular place. As such, it is always changing, for better or for worse.

The earth's atmosphere began four and a half billion years ago, as a mixture of water vapour, hydrogen, hydrogen chloride, carbon monoxide, carbon dioxide and nitrogen. By interaction with surface rocks, and from living organisms, it gradually reached its current composition. The most important part of this transformation was the conversion of much of the carbon dioxide into oxygen by abundant plant life, particularly during the Carboniferous period, when most of our coal and oil deposits were formed.

Besides atmospheric changes, the earth has passed through many temperature fluctuations. The cold periods in the more recent epochs have been termed ice ages

The United Nations Framework Convention on Climate Change, in its Article 1, defines "Climate Change" as "a change of climate which is attributed directly or indirectly to human activity that alters the composition of the global atmosphere and which is in addition to natural climate variability observed over comparable time periods"[1].

It is amazing that many Governments have endorsed this absurd and confused statement. "Climate Change" is what is "attributed" to human activity, but "Change of Climate" can take place without humans. "Natural climate variability" is not "climate change" but, apparently, is a "change of climate". There is no mention at all of "unnatural climate variability".

The IPCC[1] appears to retreat from this definition by the statement:
> "*Climate change in IPCC usage refers to any change in climate over time, whether due to natural variability or as the result of human activity*"

They seem to regard "Climate change" as identical to "change of climate", but it is unclear whether "change" and "variability" are the same thing.

This definition does not seem to inhibit the IPCC from claiming that an "unprecedented" climate change must be "anthropogenic" (jargon for "the result of human activity").

The IPCC[2] confuses matters still further by defining "Climate Change" in their "Glossary of Terms" as follows:
> "*Climate Change refers to a statistically significant variation in either the mean state of the climate or in its variability, persisting for an extended period (typically decades or longer)*"

They balk at an actual definition, preferring to state what "climate change" "refers to". "Climate" is already "usually defined" as "the average weather"[2], but now we have "the mean state of the climate", in other words, an average of an average. "Climate Change" by the IPCC definition now includes not only "change of climate", but also "its variability", both natural and unnatural. "Change" is now the same as "variation" but seems to be distinguished from "variability". We are also involved in studying the "variation" of "variability".

Chapter 2 of *Climate Change 01* is entitled "Observed Climate Variability and Change" as if the two are separate from one another. "Climate Variability:" is defined in the Glossary of Terms[2] as follows:
> "*Climate variability refers to variations in the mean state and other statistics (such as standard deviations, the occurrence of extremes etc) of the climate on all temporal and spatial scales beyond that of individual weather events*"

So, how can you tell the difference between "climate change" and "climate variability" if the latter "refers to: "all temporal and spatial scales". Of course, you cannot. Any "climate change" can be a "variability" if you take a long enough "temporal scale". Yet the IPCC tries to make such a distinction.

There is a general implication by IPCC spokespersons and by politicians that "Climate Change" is unusual, always harmful., invariably caused by humans, and must be stopped at all costs. Yet "climate change" is inevitable, unstoppable, routine, and something to be endured, for better or for worse, like death and taxes. And, if you wait long enough, it is only a form of "climate variability".

References
1 *Climate Change 01.* "Summary for Policymakers", footnote 1 to page 2
2 *Climate Change 01.* "Glossary." Page 788

3. Global Warming. What Evidence?

The most often cited evidence that the surface of the earth is warming is the global record resulting from a combination of many weather station and ship measurements. Figure 3.1 gives the latest version supplied by the Climate Research Unit, University of East Anglia[1]

Figure 3.1 is reproduced, in one form or another, no less than nine times in Chapter 2 "Observed Climate Variability and Change" of *Climate Change 01*[2]. It is also reproduced frequently in press articles about global warming. It is therefore important to discuss in some detail whether its claim to represent the mean surface temperature of the earth can be justified.

Temperature records of local weather were established soon after the invention of thermometers and temperature scales in the eighteenth century. Early thermometers were unreliable and difficult to calibrate[3,4,5]. Liquid-in-glass thermometers require a capillary tube of uniform diameter and a clearly divided scale. Glass is a cooled liquid which slowly contracts over time. Liquid-in-glass thermometers therefore read high if they are not frequently calibrated. This is true even of modern thermometers with improved low-shrinkage glass.

The earlier measurements, up to 1900 or so, would have been made on thermometers calibrated in single degrees (usually Fahrenheit), made from ordinary glass. One possible reason for the rise in temperature shown between 1910 and 1940 is the difficulties of calibration of thermometers in remote parts of the world during the two world wars.

The early measurements were made mainly near large towns in the Northern Hemisphere. Even today, measurements are not available for many regions of the earth's surface, particularly those remote from cities and buildings and for most of the oceans.

The instrumentation used to measure temperature in weather stations has hardly changed for over a century. Figure 3.2 shows the equipment currently in use at the airport on the Isle of Man On the left is the apparatus for measuring relative humidity, dependent on the properties of human hair, invented by Benjamin Franklin in the late 18th century. Then in the middle is Six's maximum and minimum thermometer and the wet and dry bulb thermometers which go back to the same era. The equipment is contained in a Stevenson's screen, invented by Robert Stevenson, the great lighthouse engineer and father of the author Robert Louis Stevenson, in the early part of the 19th century. At that time thermometers were often placed in direct sunlight, or on the walls of buildings. Change to the general use of the Stevenson screen took many years.

The screen is painted white, intended to minimise heating by the sun, or loss of heat at night to the atmosphere. However, a white painted surface is not a perfect reflector or a zero emitter. White paint absorbs 30 to 50% of the sun's radiation, so solar heat contributes to

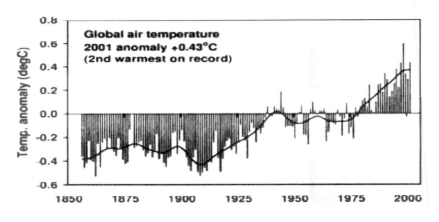

Figure 3.1   The global surface air temperature record, as compiled by the University of East Anglia[1]

the temperature measured inside the screen, particularly on a still day with little air circulation. It will increase with time as the paint deteriorates or gets dust on it, or the louvres develop cobwebs. The emissivity of a white painted surface is 85 to 95%, almost as high as black paint. On a still cloudless night the box will cool below the air temperature and influence the thermometers.

The air entering the screen will have exchanged heat with any neighbouring buildings, roads, vehicles, aircraft. It will be affected by the locality, urban or rural, and by any shelter around the site. All these properties can change over time, so influencing the measurements. Most of the changes will tend to increase the measured temperature.

In order to read the instruments, the door must be opened, so changing the air properties within the box. A change to the use of thermistors, which has taken place recently in some stations will alter this. A change to automatic recording will inevitably increase the measured temperature since there will no longer be a need to open the box to read the instruments.

Figure 3.2   Equipment for weather monitoring currently in use at the airport, the Isle of Man

Temperature records in a particular locality will not agree with those from other localities for a whole variety of reasons: elevation, proximity to a coast, wind conditions, and so on. Also weather stations come and go. There are very few with an uninterrupted record for very many years. In order to obtain a combined temperature record for many stations the procedure described by Hansen and Lebedeff[6] is used. A Mercator map of the world is divided into latitude/ longitude squares. Figure 3.1 employed 5°x5° squares. Then for each month of each year in the time sequence, acceptable weather station records are identified, and a mean of their monthly averages calculated. This average is then subtracted from the average of the means of the records for the same month in the same square, over a reference period (currently 1960–1990). The result is the *temperature anomaly* for that month. Figure 3.1 plots annual, globally averaged, temperature anomalies, calculated in this way, not individual or averaged temperature readings as such.

The intermittent changes in the record (Figure 3.1), and the irregular behaviour of neighbouring 5° x 5° grid boxes[7] is inconsistent with a steady global temperature influence, such as may come as a result of changes in the atmosphere, and points to mainly local, surface effects as their cause. The rise in the combined temperature anomalies (Figure 3.1), from 1910 to 1940 cannot be explained as a result of the greenhouse effect, since emissions were low at that time. The fall in temperature from 1940 to 1975 is inconsistent with the rise in emissions over this period.

The use of annual temperature anomaly averages in 5° x 5° squares is illustrated in Figure 3.3[8] which shows the temperature changes for the winter months of December, January and February between the years 1976 to 2000 for each square. The amounts of change are indicated by the size of the red dots (for a rise) and blue dots (for a fall). They show the grids where suitable measurements are currently available. The figure also shows that the rise in global temperature from 1976 to 2000, as indicated in Figure 3.1, was largely due to rises in temperature of weather stations in the USA, Northern Europe and the former Soviet Union, for the winter months. There was no temperature rise over this period for weather stations in the Arctic, or Antarctic, and only minimal rises for the Southern Hemisphere, or for the oceans.

The reason for the rise in temperature shown by the global surface temperature record (Figure 3.1) over the years 1976 to 2000 was therefore mainly due to improved winter heating conditions around land-based weather stations in the Northern Hemisphere.

The assumption that a temperature record from a city or an airport can be considered to represent temperature behaviour of a surrounding forest, farmland, mountain area, or desert is absurd. Table 3.1 shows that most of the energy given off by combustion of fossil fuels is given off in the neighbourhood of urban areas

The mean energy emitted by combustion of fossil fuels over the whole world is 0.02 Watts per square metre. However, this energy is emitted in a highly irregular manner. Over the USA the mean figure is 0.31 W/sq. m, and over California 0.81 W/sq. m. If energy emission is assumed proportional to population density, then the figure for San Francisco is 89.24 W/sq. m. For Germany the average is 1.23 W/sq. m., and for the industrial area of Essen, 221.65 W/sq. m. For New Zealand the average is 0.8W/sq. m., and for the city of Auckland, 28.2 W/sq. m.

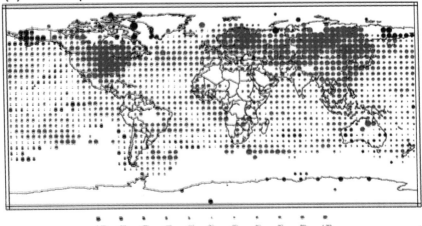

Figure 3.3   Changes in the combined weather station and ship temperatures for the winter months December, January and February for the years 1976 to 2000[8]. Size of dots shows the amount of change; red dots, a rise, blue dots, a fall

These figures should be compared with the claimed "global warming" made by the IPCC[10] for the build-up of greenhouse gases in the atmosphere since the year 1750: the amount of 2.45W/sq. m. It is clear that the predominant location of weather stations close to cities and airports can lead to "global warming" from local energy emissions which can considerably exceed the claimed effects of greenhouse gases.

Figure 3.1 violates a basic principle of mathematical statistics which asserts that a fair average of any quantity cannot be made without a representative sample. Table 3.2 shows the approximate distribution of climate zones on the earth's surface. Weather stations are situated almost entirely in the "urban area" category, only 1% of the earth's surface, where energy emission is many times the amount claimed to be caused by the greenhouse effect.

The comparison between temperature measurements made in regions remote from human habitation and those by weather stations has been displayed dramatically by Mann and Bradley[12,13], and promoted by the IPCC[14] as shown in Figure 3.4.

Figure 3.4 has several interesting features. The blue curve represents an amalgamation of "proxy" temperature measurements which mainly involve deductions from the width of tree rings, which are highly inaccurate since tree rings only show growth in the summer, so indicate only summer temperatures. The representativity of the samples is even worse than with the weather station data, but at least the measurements are all far from human activity.

TABLE 3.1 Energy Emission Statistics[9]

Locality	Mean Energy Emission in Watts per square metre
World	0.02
USA	0.31
California	0.81
San Francisco	89.24
Germany	1.23
Essen	221.65
New Zealand	0.08
Auckland	28.2
Claimed "Greenhouse Effect" since 1750	2.45

Mann and Bradley[12,13], and *Climate Change 01*[14], claim that Figure 3.4 proves that the weather station measurements are influenced by "anthropogenic" factors. Of course, this is probably true. But they fail to recognise that the "anthropogenic" effect is caused by local energy emissions around weather stations, not by changes in the atmosphere above them.

The blue curve, for proxy measurements, shows a recent increase, though within the error estimates from past measurements. Some of this increase can be attributed to enhanced tree growth from the increased carbon dioxide.

Recent proxy results do not always confirm a warming trend. As an example, see Figure 3.5. which shows temperatures derived from tree ring measurements from Northern Siberia[15].

Figure 3.5, in contrast to Figure 3.4, shows the "medieval warm period", 900 to 1100, the "little ice age" in the early 1800s, and the large peak in the 1940s which is also visible in Figure 3.1.

Many weather station measurements from remote areas, apparently uninfluenced by changes in the surroundings, also show no evidence of "global warming"[16,17].

The three authorities responsible for the combined surface record The University of East Anglia, The Goddard Institute for Space Studies, and The Global Historical Climate Network, (GHCN) have recognised that records close to towns are subject to "urbanisation" effects, and they have claimed to have applied "corrections" to the data, such as those displayed in Figure 3.1.

TABLE 3.2 Climate Zones on the earth's surface[11]

Category	Percentage
Ocean	70.8
Arable land	2.9
Permanent pasture	7.6
Forest and Woodland	9.0
Mountains	4.0.
Desert	2.0
Ice	2.0
Urban areas	1.0
Miscellaneous	0.7

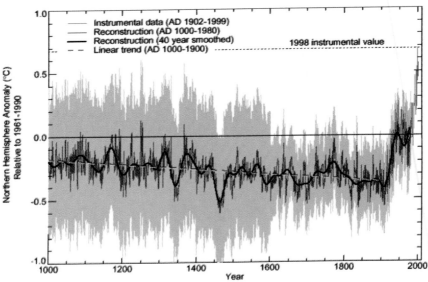

Figure 3.4 Comparison of Northern Hemisphere temperature record from proxy measurements (in blue) with weather station measurements (in red): from Mann and Bradley[12,13] and[14]. Note the additional temperature rise from proximity of weather stations to urban areas. The gray region represents an estimated 95% confidence interval.

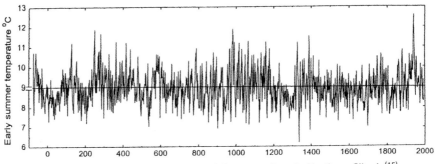

Figure 3.5 "Proxy" temperatures deduced from tree rings in Northern Siberia[15]

The procedure is to compare a record from an urban area with one in the same district which is rural, and correct the urban record according to the difference between them. Hansen et al[18] have recently identified "rural" stations that are shown to be "unlit" at night. For the USA this reduces "rural" stations to only 250, which are rather far apart for adequate correction. The authors admit that this system cannot be applied to the rest of the world.

To begin with, it can only be done where suitable records for comparison are available over a reasonable length of time. This means the corrections cannot be made where records are sparse, which applies to most of the globe, and they cannot be applied for recent records which have not been going long enough. Lists of records that have been "corrected", are only available from Hansen[19], and the "corrections" seem to be very small. The "corrections" claimed by the University of East Anglia do not seem to exist[20].

The most serious defect of the method is that it assumes that there is no "urbanisation" effect for the "rural" record that is taken as a standard. Many studies[4] have shown that there is an "urbanisation" effect even at stations that are far from cities, or near cities with a small local population. Surrounding buildings, roads, concrete, vehicles, aircraft, and increasing shelter can all provide an upwards bias even to "rural" records.

It should not really be possible to "correct" an unrepresentative sample, but the best prospects would be with records that are extensive in coverage and in time, under the same national administration, and of known high quality. The only records that can qualify are those for the continental United States. It is therefore significant that the corrected mean surface record for the United States (Figure 3.6) shows no evidence of overall "global warming" since 1920, and the slight increase over the previous period is probably due to increased energy exposure of "rural" weather stations[18].

Figure 3.6 shows a temperature rise from 1910 to 1940 which fell again from 1940 to 1975. The most plausible explanation for this is the growth of towns in the first period, and the move of weather equipment to airports in the second.

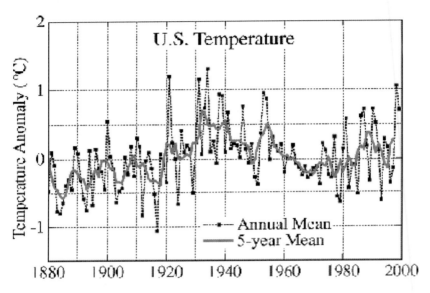

Figure 3.6   Temperature record for the continental United States[19]

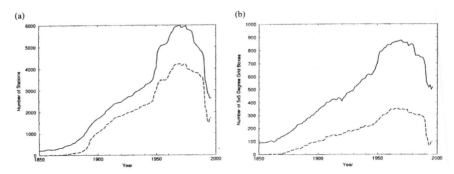

Figure 3.7   Graphs showing (a) Numbers of temperature records available (upper curve) and Maximum and Minimum records available (lower curve). (b) Number of 5° x 5° grids available with temperature records (upper curve) and Maximum and Minimum temperature (Lower curve)[21]

Another factor that has influenced the weather station record (Figures 3.1, 3.6), is the variable number of stations available. Figure 3.7[21] shows how station numbers and available grids have varied. The large increase after 1950 was an increase in rural stations, and of stations at airports, and it partly accounts for the fall in the combined temperature from 1950 to 1975 shown in Figure 3.1. The wholesale closure of mainly rural stations in 1989, combined with the increased energy release at airports partly accounts for the increase in the combined temperature record shown in Figure 3.1 since 1989.

Then, there is the question of sea surface records. Figure 3.1 claims to incorporate sea surface records. As the ocean is 70.8 percent of the earth's surface, inclusion of sea surface temperature records is rather vital for representativity.

Sea surface temperature records are voluminous (80 million observations) and extensive, but they suffer from serious defects. Weather station records, plus the limited numbers of fixed buoys, have temperatures taken in the same place over a period of time by qualified staff and continuity of administration. Ship measurements are rarely in the same place, the procedures are far from standard, and the staff and control are often less than professional. Most early measurements were by recovery of samples by buckets, and more recent measurement at the engine intake. There are also measurements of night marine air temperature, from deck measurements, usually from a Stevenson screen

Folland and Parker[22] suggested corrections to sea surface temperature measurements which have led to their amalgamation with the land-based measurements by British workers to give the combined record of Figure 3.1. One persistent problem is incomplete information; for example, what sort of bucket was used to collect a sample. When in doubt, Folland and Parker allow the sea surface temperature to coincide with the land-based temperature, which means that there was little obvious change from the addition of the sea surface data.

Doubts that the use of sea surface data is justified have recently been shown by Parker himself, in association with Christy and others[23] who found that there is a discrepancy between the current measurement of sea surface temperature by engine intakes, and the measurements closer to the surface made by fixed buoys and on ship's decks.

Figure 3.8[24] shows a comparison between the accepted surface/sea surface combined measurements, those modified by the correction of Christy et al[23] to incorporate only marine air temperatures, and the globally averaged satellite measurements. It shows the large discrepancy between sea surface and engine intake measurements, and the corresponding lack of overall increase shown by the satellite measurements (see below).

The United States compilers of global temperature[21,25] refuse to recognise the sea surface data from ships as reliable, so that their compilations deal only with land-based records, plus recent sea surface data from satellites.

Recent satellite measurements of mean sea surface temperature[26] have found no evidence of distinguishable warming for the past 16 years, after allowance is made for the effects of volcanic eruptions and El Niño events.

The mere presence of a ship in the ocean is bound to influence

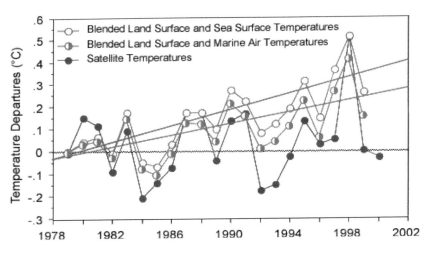

Figure 3.8   Comparison between blended land surface and sea surface temperatures, blended land surface and marine air temperatures and satellite temperatures, from 1978[24]

the temperature of the surrounding sea surface temperature, so that ship measurements are subject to the same influences of greater size and energy consumption in ships as are the land measurements. As for the night marine air temperature, where do they place the Stevenson screen? Usually up against the funnel.

Levitus et al[27] have published estimates of the heat content of the world's oceans since 1948 (Figure 3.9) which show a significant rise with time. As usual, there is a doubt as to whether measurements covered a representative selection of the world's oceans, and the use of only one standard error for the error bars, instead of the usual two standard errors (indicating 95% confidence) means that there is a one in three chance that the figures fall outside them. Also, the removal of the influence of the exceptionally vigorous El Niño event of 1997-1998 would make the change less impressive. Even if there was a genuine increase in ocean heat, it could not have been related to temperature changes in the atmosphere, as measured by satellites, or to satellite measurements of sea surface temperature, which were negligible since 1984[26], if corrections are made for El Niño.

Since the temperature rise shown in Figure 3.1 between 1910 and 1945 could not have been due to greenhouse gas increase, the question is, what was the cause? Figure 3.10 shows changes in individual 5° x 5° grids over this period[28]. As before, the size of the dots in individual grids gives the size of the change, red for a rise and blue for a fall. Most of the temperature rises between 1910 and 1940 (shown by larger red dots) were in the United States and in the Atlantic ocean. The US rise is probably a result of the increased size of American cities when Europe was affected by war. The ocean measurements would have been affected by the absence of lights on ships during the two wars, which meant that measurements had to be made below deck, plus great increases in the size and energy consumption of ships.

It is interesting that some of the greatest increases over the period took place in the Arctic, whereas the Arctic stations showed a fall in temperature over the period 1901 to 2000[7] The unrepresentative character of the coverage is evident from Figure 3.10, so the apparent trend cannot be taken too seriously.

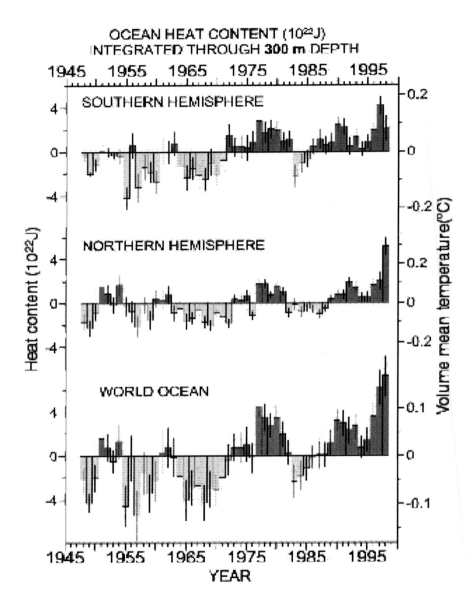

Figure 3.9 Ocean heat since 1948, according to Levitus et al[27]. Error bars are only one standard error

To summarise this section:
- The combined surface record (Figure 3.1 and its companions) is not a reliable indicator of global temperature change.
- It is based on a heavily biased sample, so that it actually represents a modified version of the temperature conditions surrounding weather stations, as influenced by larger than average energy emissions, towns, buildings, roads, vehicles and aircraft. The increase of a mere 0.6°C over 140 years could easily represent changes in surrounding energy conditions.
- The large differences often shown between neighbouring grids on the temperature maps[7] indicate that they are locally influenced.
- The fact that many more remote stations and most "proxy' measurements show no overall warming, indicate the absence of a steady warming trend over the past century.

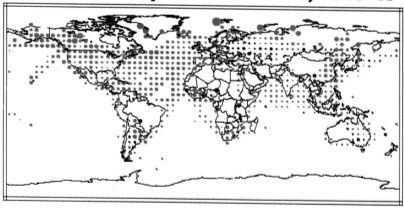

Figure 3.10   Temperature change for individual 5° x5° grids from 1910 to 1945. The area of the dot shows size of increase, per decade (red dots) or decrease per decade (blue dots)[28]

Measurements of temperature in the lower troposphere confirm this conclusion.

Since 1958 there have been temperature measurements by weather balloons (radiosondes) from 63 sites which were scattered over the globe, but were mainly over land, in the more highly populated parts of the world[29].

The mean global temperature for the lower troposphere is shown in Figure 3.11[29].

Figure 3.11 shows that the temperature of the lower troposphere was approximately constant since 1977, in contrast to the combined surface measurements (Figure 3.1). The annual fluctuations shown by Figure 3.11, are very similar to those in Figure 3.1, suggesting that both methods faithfully record temperature changes due to volcanoes, solar fluctuations and ocean variability.

The sudden jump which took place between 1955–1975 and 1976–1999 is difficult to explain. It could be a consequence of the limited coverage, or changes in instrumentation, or it could be a genuine climate adjustment. There is, however, no justification in putting a linear regression line through the whole series, as there is no

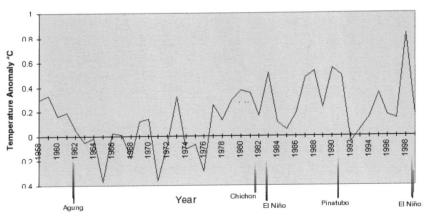

Figure 3.11   Mean global temperature of the lower troposphere from 63 radiosondes[29]

evidence of a regular linear change. After all, the figure for 1958 was the same as that for 1999.

Since 1979, NOAA satellites have been measuring the temperature of the lower troposphere using Microwave Sounder Units (MSUs) The method is to measure the microwave spectrum of atmospheric oxygen, a quantity dependent on temperature. It is much more accurate than all the other measures and, also in contrast to other measurements, it gives a genuine average of temperature over the entire earth's surface. Various efforts to detect errors have not altered the figures to any important degree. The record is shown in Figure 3.12[30].

The annual fluctuations in this record agree well with those in the combined surface record (Figure 3.1) and with the radiosonde record (Figure 3.11). However, when these fluctuations are removed from this record[31] there is no overall evidence of a warming trend. The very large effect of the El Niño event in 1998 gives a spurious impression of a small upwards trend.

The absence of a distinguishable change in temperature in the lower troposphere over a period of 23 years, is a fatal blow to the "greenhouse" theory, which postulates that any global temperature change would be primarily evident in the lower troposphere. If there is no perceptible temperature change in the lower troposphere then the greenhouse effect cannot be held responsible for any climate change on the earth's surface. Changes in precipitation, hurricanes, ocean circulation and lower temperature, alterations in the ice shelf, retreat of glaciers, decline of corals, simply cannot be attributed to the greenhouse effect if there is no greenhouse effect to be registered in the place it is supposed to take place, the lower troposphere.

By contrast, the combined surface record (Figure 3.1) is nothing more than a record of an averaged presentation of the temperature conditions near weather stations and ships, influenced by the additional energy emissions associated with urban environments, and not a valid record of mean global temperature because it uses a highly biased sample[32].

We therefore have a situation where all direct measurements of mean global temperature show no evidence of a perceptible change, at least over the past 23 years, and probably over the past century. There is therefore no reason for any precautionary measures intended to prevent or limit such a temperature change.

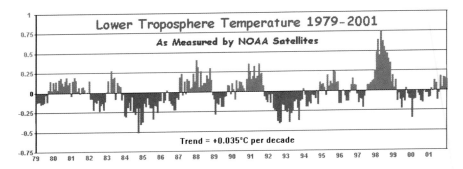

Figure 3.12   Mean global temperature anomalies of the lower troposphere, as measured by MSU units (NOAA)[29]

References

1. Climatic Research Unit, University of East Anglia, UK, http://www.cru.uea.ac.uk
2. *Climate Change 01* Chapter 2, Figures 2.1, 2.4, 2.5, 2.6, 2.7, 2.8, 2.12, 2.19, 2.21.
3. McConnell, A, 1992. "Assessing the value of historical temperature measurements" *Endeavour* 16 (2) 80-84.
4. Gray, V R. 2000. "The Surface Temperature Record" http://www.john-daly.com. Guest contributions.
5. Daly, J L. 2000. "The Surface Record : Global Mean Temperature and how it is determined at the surface level". http://www.greeningearthsociety.org
6. Hansen, J and S Lebedeff. 1987. "Global Trends of Measured Surface Air Temperature" *Journal of Geophysical Research* 92 13345-13372.
7. *Climate Change 01* Chapter 2, "Observed Climate Variability and Change", Figures 2.9 and 2.10, pages 116-7.
8. *Climate Change 01* Chapter 2 "Observed Climate Variability and Change" Figure 2.10a, page 117.

9   Carbon Dioxide Information and Analysis Center, Oakridge, Tennessee, http:/cdiac.esd.ornl.gov/trends, *United Nations Energy Handbook* and estimates.
10 *Climate Change 01* Chapter 6. "Radiative Forcing of Climate Change" Figure 6.6, page 392.
11 Adapted from the CIA Factbook. http://www.oda/gov/cia/publications/factbook
12 Mann, M E, R S Bradley & M K Hughes. 1998. "Global-scale temperature patterns and climate forcing over the past centuries *Nature* 392 778–787.
13 Mann, M E, R S Bradley & M K Hughes. 1999. "Northern Hemisphere Temperatures During the Past Millennium: Inferences, Uncertainties and Limitations" *Geophysical Research Letters* 26 759–76.
14 *Climate Change 01* "Summary for Policymakers", Fig 1. Page 3.
15 Naurzbaev, K M, & E A Vaganov. 2000. "Variation of early summer and annual temperature in East Taymir and Putoran (Siberia) over the last two millennia inferred from tree rings" *Journal of Geophysical Research* 105 7317–7326.
16 Daly, J L. 2001. "What the stations say". http://www.john-daly.com
17 CO2 Science Magazine. 2001 http://www.co2science.org
18 Hansen, J, R Ruedy, M Sato, M Imhoff, W Lawrence, D Easterling, T Peterson, & T Karl. 2001. "A closer look at United States and global surface temperature change" *J Geophys Research* 106, 23,947–23,963.
19 Hansen, J et al. 2001. http://www.giss.nasa.gov/data/update/gistemp
20 Hughes, W S. 2001. http://www.webace.com.au/~wsh
21 Peterson, T C, & R S Vose. 1997. "An Overview of the Global Historical Climatology Network Temperature Database" *Bulletin of the American Meteorological Society* 78 2837–2849.
22 Folland, C K & D E Parker. 1995. "Correction of instrumental biases in historical sea surface temperature data" *Quarterly Journal Meteorological Society* 121 319–367.
23 Christy, J R, D E Parker, S J Brown, I Macadam, M Stendel & W B Norris. 2001. "Differential Trends in Tropical Sea Surface and Atmospheric Temperatures since 1979" *Geophysical Research Letters* 28 183–186.
24 *World Climate Report.* 2001. 6 (9) "Satellite 'Warming' vanishes".
25 Goddard Institute of Space Studies (GISS) at http://www.giss.nasa.gov and GHCN at http://www.ncdc.noaa.gov

26 Strong, A E, E J Kearns & K K Gjovig. 2000. "Sea Surface Temperature Signals from Satellites – An Update" *Geophysical Research Letters* 27 1667–1670.
27 Levitus, S, J Antonov, T P Boyer & C Stephens. 2000. *Science* 287 2225–2229.
28 *Climate Change 01* Chapter 2 Figure 2.9b page 116.
29 MSU Homepage. http://wwwghcc.msfc.nasa.gov/MSU/msusci.html
30 "Still Waiting for the Greenhouse" http://www.john-daly.com
31 Michaels, P J & P C Knappenberger. 2000. "Natural Signals in the MSU lower tropospheric temperature records" *Geophysical Research Letters* 27 2905–2908.
32 Gray, V R. 2000. "The Cause of Global Warming." *Energy & Environment* 11 613–629.

4. Greenhouse Gases and Aerosols

The Greenhouse Effect

The earth's temperature today is modified by the presence in the atmosphere of several gases which are capable of absorbing part of the infrared radiation that is given off by the earth, and re-radiating part of it back again (Figure 1.2). Without these gases the mean surface temperature could be 33°C less than it is today. The most important gases are water vapour, carbon dioxide, methane, ozone, halocarbons and nitrous oxide. In addition to these gases there are similar effects due to aerosols, amongst which are ordinary clouds, dust, and carbon (soot).

Any changes in these gases and aerosols will alter the thermal balance in the atmosphere, and so the mean temperature of the earth. The recent increase in the concentration of the minor greenhouse gases has led to the belief that there has resulted an increased mean global temperature, the "enhanced greenhouse effect". As has been shown in the previous chapter, however, there is no evidence, so far, that this is happening. The extra radiation said to be due to the enhanced greenhouse effect is called "radiative forcing" by the IPCC.

Water Vapour

Tyndall[1] regarded water vapour as the dominant greenhouse gas. He wrote that water vapour "acts more energetically upon the terrestrial rays than upon the solar rays; hence, its tendency is to preserve to the earth a portion of heat which would otherwise be radiated into space". Baliunas and Soon[2] recently calculated that water vapour is responsible for about 88% of the absorption of infrared radiation from the earth in the 4 to 60 micron wavelength range.

As the main greenhouse gas it is unlikely that there will be any evidence of an "enhanced greenhouse effect" or of "radiative forcing" unless it can be shown that the amount and distribution of water vapour in the atmosphere is changing in such a fashion as to increase the greenhouse effect. Unfortunately there is no such record. Water vapour concentration in the atmosphere is highly variable over several orders of magnitude both locally, temporally and with height. There are simply no past or present records which could help to decide whether its radiation properties have changed or are changing. There are a number of papers which have found evidence from satellites of recent increases in water vapour in the upper atmosphere[3,4]. However, Lindzen[5,6] has pointed out that it does not follow that this effect pervades the atmosphere, or is related to the behaviour at the surface, which we have seen in Chapter 3, has no firm evidence for a warming. Rind[7] in discussing Lindzen's objections, concedes that

> "Unfortunately we do not have the observations in place to be able to do so [give a proper figure for the water vapour feedback] and it is not clear when we will"

Climate Change 01[8] says:

> Raval and Ramanathan's results[3] are difficult to interpret as they involve the effects of circulation changes as well as direct thermodynamic control"

They make similar reservations about the results of Inamdar and Ramanathan[9]

> "it still cannot be taken as a direct test of the feedback as the circulation fluctuates in a different way over the seasonal cycle than it does in response to the doubling of CO_2"[8]

The mere existence of water vapour as a greenhouse gas is simply omitted from Chapter 4 "Atmospheric Chemistry and Greenhouse Gases" of *Climate Change 01*[10] It does not include tropospheric water vapour as part of "atmospheric chemistry". The Chapter[11] begins its "Executive Summary" with

> "Currently, tropospheric O_3 is the most important greenhouse gas after CO_2 and CH_4". Water vapour does not exist!

Chapter 6 "Radiative Forcing of Climate Change"[12] also fails to mention water vapour.

The excuse usually given for this neglect is that water vapour is relegated to the status of a "feedback" where its unknown effects can be guessed in the form of a parameter proportional to the calculated temperature.

A possible theoretical treatment is the Clapeyron equation

$$dp/dT = (\lambda/T(v_2 - v_1)$$

Where p is the vapour pressure, λ is the latent heat of evaporation, T is the absolute temperature, v_1 is the volume of unit quantity of liquid, v_2 is the volume of unit quantity of vapour.

The equation represents the behaviour at equilibrium, a condition never reached in the atmosphere.

The equation was modified by Clausius to give the Clausius-Clapeyron equation, by using the Gas Law, pv=RT (R is the gas constant) to replace v_2 and ignoring v_1 altogether.

$$dp/dT = p\lambda/RT^2$$

which deals only with vapour pressure, but in addition to assuming that the atmosphere is in equilibrium (which it is not) also assumes that water vapour is an ideal gas (which it is not).

Climate Change 01[13] downplays the importance of this equation, as follows :

"In the SAR [*The Second Assessment Report*], a crude distinction was made between the effect of "upper-tropospheric" and "lower-tropospheric" water vapour, and it was implied that lower tropospheric water vapour feedback was a straightforward consequence of the Clausius-Clapeyron relation. It is now appreciated that it is only in the boundary layer that the control of water vapour by Clausius-Clapeyron can be regarded as straightforward"

At first glance, it seems reasonable that the vapour pressure of water ought always to increase with temperature, even when not in equilibrium, although Lindzen has argued that this is not necessarily so[6].

But when it is appreciated that part of the water vapour is involved in the formation of clouds, it no longer seems so reasonable.

Clouds

Along with our ignorance on the historical and current concentration and distribution of water vapour, we are equally ignorant about the past or even the present behaviour of clouds. Cloud formation is extremely complex, and the radiative behaviour of clouds depends on the type of cloud and its location. This unpredictable and unknown behaviour is, of course, dealt with by guesswork, an assumption of cloud "feedback". Although water vapour guesses in climate models are relatively uniform, those for clouds represent the greatest source of the difference between climate models, as is shown by various intercomparison studies[14,15,16,17]. There is evidence that "cloud cover" (however measured) has increased over the years[18] but there is no information on the different types of cloud. So maybe any temperature increase is nullified by increased clouds?

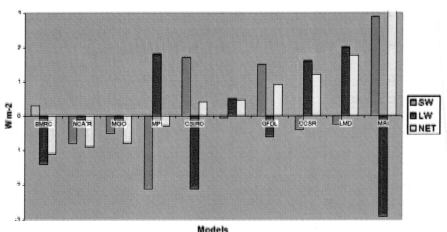

Figure 4.1   Assumptions for cloud feedback (in watts per square centimetre) made by 10 different models. Short wave (SW, terrestrial, in blue), Long wave (LW, solar, in red) and NET (in yellow; can they add up?) are separately indicated. Complete guesswork?[19]

Figure 4.1 provides proof that the modellists simply have no idea what contribution clouds make to radiative forcing. It is Figure 7.1 from *Climate Change 01*,[19] and it shows the assumptions made for the radiative forcing contributions of clouds to the radiative forcing at the top of the atmosphere of 10 models. The contributions of short wave (SW, solar) and long wave (LW terrestrial) are separately indicated. The NET figure is also shown, but the last bar indicates problems with arithmetic.

Comprehensive measurements of clouds by satellite have been made since 1983[20], but there is no evidence of any consistent trend.

Other Effects of Water

Precipitation, whether of rain or snow, also affects climate, and we have little reliable or representative information. *Climate Change 01*[21] identifies a recent increase in precipitation for the few, unrepresentative sites for which it has data. Snow precipitation affects behaviour of glaciers (often exclusively blamed on temperature), and the albedo (reflectivity) of some regions.

Although water, in the form of vapour, clouds and precipitation, has a dominant influence on the climate, and on the greenhouse effect, climate modellists are reduced to guesswork (called "feedback") in assessing its influence, in the absence of reliable historical data.

Carbon Dioxide
The Carbon Budget

Table 4.1, from *Climate Change 01*[22] gives the following components of the Carbon Budget, divided into decades

Emissions from the combustion of fossil fuel, gas flaring and cement are considered to be partitioned between the atmosphere, the ocean, and the land, according to the fractions shown.

It will be seen that the fraction of the emissions from fossil fuels and cement entering the atmosphere has fallen from 0.61±0.02 to 0.51±0.02 between the 1980s and the 1990s. It is thought that the amount entering the ocean has fallen 1.9±0.6 to 1.7±0.5 GtC in the period.

TABLE 4.1 Components of the Carbon Budget[22]

Budget Components	1980s, GtC	1980s Fraction	1990s GtC	1990s Fraction
Emissions (fossil fuel, cement)	5.4±0.3	1.0	6.3±0.4	1.0
Atmospheric increase	3.3±0.1	0.61±0.02	3.2±0.1	0.51±0.02
Ocean absorbed	−1.9±0.6	−0.19±0.11	−1.7±0.5	0.27±0.08
Land absorbed	−0.2±0.7	0.03±0.13	−1.4±0.7	0.22±0.11
*Land partitioned; land-use change	1.7 (0.6 to 2.5)			
Residual terrestrial sink	−1.9 (−3.8 to 0.3)			

Partitioning of the carbon dioxide absorbed by the earth between the ocean and the land has used two methods[23]. The first uses the change in atmospheric oxygen. Carbon dioxide absorbed on the land implies a simultaneous emission of oxygen, whereas absorption by the ocean does not. The second uses measurements of the relative abundance of the carbon isotopes ^{12}C and ^{13}C. C3 plants discriminate against ^{13}C and so change the ratio between the isotopes. Both these methods depend on an array of assumptions, and the accuracy is poor. It is, to begin with, difficult to believe that ocean absorption goes down with increased carbon dioxide concentration as suggested in Table 4.1.

The calculated figure for carbon dioxide absorbed by land is also highly inaccurate. There is a tendency to try to partition the land absorption into two components, *land-use change* which is considered to cause an emission, and *residual terrestrial sinks* which causes an absorption. The emission part is sometimes added to the fossil fuel figure to give a gross emission. To be logical, if you are going to add *emissions* you should also subtract *sinks* to give gross emissions. In reality it is almost impossible to identify land surface regions which either emit or absorb carbon dioxide, although much effort has gone into trying.

The large increase in land absorption from the 1980s to the 1990s must be due to greatly increased agriculture and forestry productivity. As there are only two points on a graph, it is difficult to say whether this trend will continue.

Emissions

Although carbon dioxide is an entirely minor greenhouse gas, it has been postulated that the emission of carbon dioxide by the combustion of fossil fuels can have an important influence on the climate, and therefore must be brought under control.

The countries of the world have been involved for several years now in discussing proposals for control of carbon dioxide emissions.

It is therefore something of a surprise to find that *Climate Change 01* has no Chapter dealing with carbon dioxide emissions, but only a very short paragraph in Chapter 2 (3.4.1. Emissions from Fossil Fuel Burning and Cement Production) and a single diagram (Figure 3.3) showing emissions from 1958 to the year 1999.[24]

Although the developed nations involved in the Kyoto Treaty process have methods for measuring their own emissions, the latest figure for world emissions from the Carbon Dioxide Information and Analysis Center is for the year 1998. This means that any measures to reduce emissions will require a wait of at least three years before they will find whether they have had any effect on global emissions.

Figure 4.2 gives emissions from combustion of fossil fuels, gas flaring, and cement production, since 1950[25], with estimates for 1999 and 2000 from BP[26].

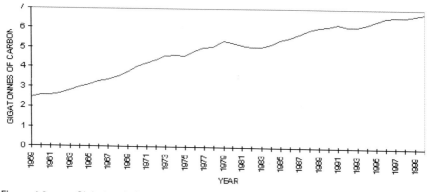

Figure 4.2   Global emissions of carbon dioxide from combustion of fossil fuels, gas flaring, and cement production[25, 26]

It is evident that carbon emissions are somewhat irregular, but that there was a fairly linear rate of increase of 0.14 GtC/yr from 1959 to 1979 which was replaced by a fairly linear rate of 0.08 GtC/yr from 1979 to 2000, almost half the previous rate. The difference is attributed (Table 1) to a considerable increase in absorption by the land surface, a change which is largely unexplained.

Atmospheric Concentrations

Carbon dioxide is one of the most difficult gases to measure. It is only in the last 20 years or so that an effective infrared analysis method has evolved, free from the necessity of cross calibration.

A half-forgotten figure is G S Callendar. He was an advocate of the existence of the greenhouse effect in the late 30s and he backed up his claims with measurements of the carbon dioxide content of the

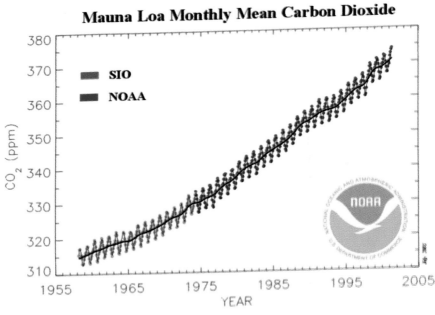

Figure 4.3   Monthly atmospheric carbon dioxide concentrations as measured near the Mauna Loa volcano, Hawaii, from 1959 to 2000[32]

atmosphere which have been discredited.. He claimed that the 19th century value of 274 parts per million had increased to 325 parts per million by 1935 as a result of the combustion of fossil fuels, and that this caused an increase in global temperature of 0.33°C[27,28,29], from Jaworowski[30].

The whole subject was revived by the measurements of Keeling[31] near the summit of the active volcano Mauna Loa on the island of Hawaii. Figure 4.3 gives the latest version of measurements at Mauna Loa[31] which combines Keeling's figures before May 1974 with those of NOAA afterwards.

It is immediately apparent that the approximately linear rate of increase between 1959 and 1974 (0.9ppm/yr) was followed by the approximately linear rate of increase from 1974 to 2000 (1.53ppm/yr) which shows no signs of changing.

Since 1969 a world-wide network of stations measuring atmospheric concentrations of carbon dioxide has been established. In contrast to the weather stations measuring temperature, which are predominantly close to cities and buildings, the carbon dioxide network is deliberately located at places that are "remote from fossil fuel combustion and vigorous plant activity"[32]. In other words, once more, they do not provide a fair average, and they do not provide information on the carbon dioxide concentration above most land surfaces, where most concerns are expressed about the consequences of the greenhouse effect. The claim that carbon dioxide is a "well-mixed" gas are patently untrue, as most stations avoid measurements when the wind is in the wrong direction (from land) and record only concentrations from the sea.

It is postulated that changes in atmospheric carbon dioxide concentration are caused by combustion of fossil fuels. If we assume, for the moment, that the Mauna Loa record can be considered to represent globally averaged carbon dioxide concentrations, then each annual increment in atmospheric addition should be the result of the annual emission of carbon dioxide. The calculation (Figure 4.4) shows, that generally the atmospheric increment is less than the emission. The difference must surely represent absorption by the earth.

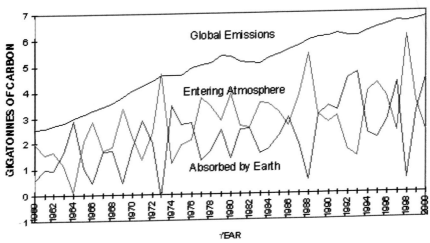

Figure 4.4   Global Emissions of carbon dioxide from combustion of fossil fuels and cement between 1960 and 2000, compared with the annual increment into the atmosphere, and, by difference, the amount absorbed by the earth

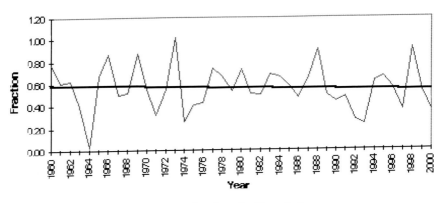

Figure 4.5   The Airborne Fraction of carbon dioxide emissions

There does not seem to be a close relationship between the annual increment of carbon dioxide emitted by combustion of fossil fuel, cement production, and gas flaring, and the increment measured in the atmosphere. In 1966, 1969, 1973, 1988 and 1996 all, or almost all emissions entered the atmosphere. These were all El Niño years. In 1964, 1992 and 1993 none, or little, entered the atmosphere. 1964 was a La Niña year and so was 1993. So the fluctuations seem to be heavily dependent on the Southern Oscillation in the ocean.

It is, however, obvious from Figure 4.4, that carbon entering the atmosphere has, on average, been constant at about 3 GtC/yr, since 1974, whereas the carbon absorbed by the earth has (when smoothed) increased from 1.5GtC/yr to 3.4GtC/yr over the same period because of increased plant production.[33]

Figure 4.5 is a plot of the Airborne Fraction, that is to say, the fraction of emissions due to combustion of fossil fuels, cement manufacture and gas flaring, that arrives in the atmosphere. It will be seen that it has fallen from 0.59 to 0.52 between 1960 and 2000.

Methane

Methane CH_4 is the third most important greenhouse gas, after water vapour and carbon dioxide. There is considerable disagreement on quantities emitted[34]. It is emitted partly from natural sources, 92–270 $TgCH_4$/yr, and partly from "anthropogenic" sources (which include agriculture), 315–350 $TgCH_4$/yr

The behaviour of methane in the atmosphere is completely different from that of carbon dioxide, since it decomposes rapidly, so that the atmospheric concentration depends on a constant source of supply. The "lifetime" of methane in the atmosphere is around 10 years[35].

The atmospheric concentration of methane has been measured at the same sites that measure carbon dioxide since 1984, so it suffers from the same defect of only measuring "background" figures. The true average, or figures over land or industrial areas are unknown. The latest globally averaged figures are shown in Figure 4.6[36].

It will be seen that the rate of increase in the atmospheric concentration of methane has been falling since 1984, and it is now negative, so that the concentration itself is now falling. Methane therefore cannot be considered at present as an important greenhouse gas, despite the emphasis placed on it by the IPCC and by Kyoto negotiators.

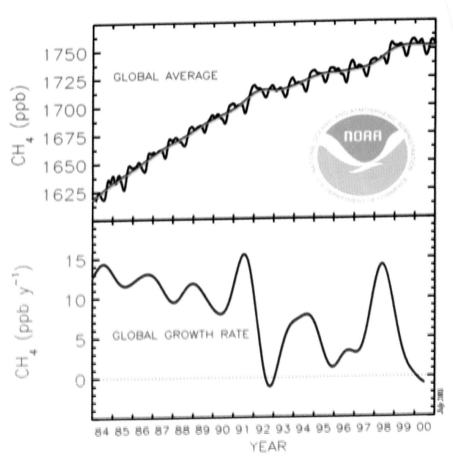

Figure 4.6   Globally averaged atmospheric concentration of methane, and its rate of change since 1984[36]. Note that the concentration is now falling

Other Greenhouse Gases

Climate Change 01[37] lists 63 greenhouse gases additional to water vapour, carbon dioxide and methane. Many of these are halocarbons which are largely being reduced because of their supposed effects on ozone in the stratosphere. Since their calculated effects are marginal they will not be further discussed here.

Aerosols

Ordinary clouds are the most important aerosols influencing the climate, yet they do not feature in the *Climate Change 01* Chapter 5 on "Aerosols: their Direct and Indirect Effects"[38]. The uncertainties connected with their properties are concealed in their status as "feedbacks". Chapter 5 does, however deal with *influences* on clouds which are dealt with separately from the influence of the clouds themselves.

The Chapter lists eight kinds of aerosol, each of which has direct and indirect effects. They are:

Soil dust, Sea Salt, Industrial dust, Carbonaceous aerosols, Primary biogenic aerosols, Sulphate, Nitrates and Volcanoes.

Most of the effects of aerosols are difficult to measure, and the uncertainties, all of which are unquantified in a scientific sense, are very great. Many aerosols cool the earth. Those predominantly due to industry should cool the Northern Hemisphere more than the Southern Hemisphere, an effect not observed, even by the most reliable temperature records.

The effects of aerosols, and their uncertainties, are such as to nullify completely the reliability of any of the climate models, as will appear evident in the next Chapter.

References

1 Tyndall, J. 1859. "Note on the transmission of heat through gaseous bodies". *Proceedings of the Royal Society of London* 10, 155–158

2 Baliunas, S, & W. Soon. 1999. "Pioneers in the Greenhouse Effect". *World Climate Report* 4 (19). At http://www.greeningearthsociety.org/climate

3. Raval, A, & V Ramanathan. 1989. "Observational determination of the greenhouse effect" *Nature* 342 758-61
4. Rind, D, et al. 1991. "Positive water vapour feedback in climate models confirmed by satellite data" *Nature* 349 500-502
5. Lindzen, R S. 1994. "Climate Dynamics and Global Change" *Annual Review* of Fluid Mechanics 26 353-378
6. Lindzen, R S. 1993. "Absence of Scientific Basis". *National Geographic Research and Exploration* 9 191-200
7. Rind, D. 1998. "Just Add Water Vapour" *Science* 281 1152-1153
8. *Climate Change 01* Chapter 7 "Physical Climate Processes and Feedbacks" page 425
9. Inamdar, A K & V Ramanathan. 1998. "Tropical and Global Scale Interactions among Water Vapor, Atmospheric Greenhouse Effect and Surface Temperature." *Journal of Geophysical Research* 103, 32177-32194
10. *Climate Change 01* Chapter 4 "Atmospheric Chemistry and Greenhouse Gases", pages 240-288
11. *Climate Change 01* Chapter 4, page 241.
12. *Climate Change 01* Chapter 6 "Radiative Forcing of Climate Change", pages 349-416
13. *Climate Change 01* Chapter 7 "Physical Climate Processes and Feedbacks", page 427
14. Cess, R D et al. 1990. "Intercomparison and Interpretation of Climate Feedback Processes in 19 Atmospheric General Circulation Models", *Journal of Geophysical Research* 95 16601-16615
15. Cess, R D et al. 1989. "Interpretation of Cloud-Climate Feedback as produced by 14 Atmospheric General Circulation Models" *Science* 245, 513-516
16. Cess, R D. 1996. "Cloud feedback in atmospheric general circulation models: an update". *Journal of Geophysical Research* 101 12791-12794.
17. Gates, W L et al. 1999. *Bull Am Met Soc* 80 29-55
18. *Climate Change 01* Chapter 2 "Observed Climate Variability and Change", Figure 2.3, page 109
19. *Climate Change 01* Chapter 7 "Physical Climate Processes and Feedbacks", Figure 7.2 page 430
20. Rossow, W. B. ISCCP. 2001. http://iscpp.giss.nasa.gov
21. *Climate Change 01* Chapter 2. Page 159

22 *Climate Change 01* Chapter 3, "The Carbon Cycle and Atmospheric Carbon Dioxide" page 190
23 *Climate Change 01* Chapter 3, Figure 3, pages 205-207
24 *Climate Change 01* Chapter 3, page 204
25 Carbon Dioxide Information and Advisory Center, http://cdiac.esd.ornl.gov/trends
26 British Petroleum. http://www.bp.com/centres/energy/world
27 Callendar, G S. 1938. *Quarterly J of the Meteorological Society* 64, 223
28 Callendar, G S. 1940. *Quarterly J of the Meteorological Society* 65 395
29 Callendar, G S. 1958. *Tellus* 10 243
30 Jaworowski, Z. 1997. *Twenty first Century* 10 (1) 42-52
31 Keeling, C D et al 1989 in D H Peterson (Ed). *Geophysical Monograph* 55 American Geophysical Union, pages 305-363, updated by Reference 25
32 Carbon Cycle Group, NOAA, http://www.cmdl.noaa.gov/ccg
33 CO2 Science magazine, http://www.co2science.org/journal
34 *Climate Change 01* Chapter 4, Table 4.2, page 250
35 *Climate Change 01* Chapter 4 Table 4.3 Page 251
36 Dlugokencky, E J, K A Masarie, P M Lang & P P Tans. 1998. "Continuing decline in the growth rate of the atmospheric methane burden." *Nature*, 393 447-450, plus update at NOAA Carbon Cycle Group 2001 http:///.cmdl.noaa.gov/ccg
37 *Climate Change 01*. Chapter 4 Table 4.1, pages 244-5
38 *Climate Change 01*. Chapter 5 "Aerosols, their Direct and Indirect Effects" pages 290-348

5. Sea Level

Climate Change 01[1] in its "Summary for Policymakers" has a headline "Global average sea level has risen" This is followed by the statement "Tide gauge data show that global average sea level rose between 0.1 and 0.2 metres during the 20th century". But how representative are tide gauge data?

Chapter 11 "Changes in Sea Level"[2] has 41 pages of speculations, estimates, forecasts and models. Only just over three pages[3] are devoted to "Mean Sea Level Changes over the Past 100 to 200 Years".

The level of the sea is a highly variable quantity; affected by the earth's orbit, by the moon, by all aspects of the climate. When it is measured from a land-based facility, the measurement is affected by changes in the level of the land.

There are many land-based locations around the world where the level of the sea is measured, and often automatically recorded. These measurements have been collected and made available to the public on the Permanent Service for Mean Sea Level (PSMSL) website[4]. Currently there are some 1850 records listed. They vary considerably in their length. They also vary in the quality of equipment and the degree of supervision.

The measurement stations are most often near large cities in the Northern Hemisphere. They, therefore, cannot be considered as providing a representative sample of the world's oceans. Any averages derived from them cannot be regarded as evidence of a globally averaged change in sea level, but only of the particular sites of measurement. *Climate Change 01*[5] puts it this way:

> "The sea level records contain significant interannual and decadal variability and long records are required in order to

> estimate reliable secular rates that will be representative of the last century. In addition, sea level change is spatially variable because of land movements and of changes in the ocean circulation. Therefore a good geographic distribution of observations is required. Neither requirement is satisfied with the current tide gauge network."

Corrections to land measurements such as the "Post Glacial Rebound" (the continued recovery from the last ice age) and plate movements have to be made by the use of models that are often based on limited geographical regions, and are prone to large uncertainties. Again, *Climate Change 01*[5]:

> "However, this procedure may underestimate the real current eustatic change because the observed geological data may themselves contain a long-term component of eustatic sea level rise."

Corrections due to urban development are even more difficult to make The level of all cities falls as ground water is removed and heavy buildings are erected. Tide gauge equipment may itself tend to subside after years of battering by the sea.

The need for corrections means that the quoted figures for sea level change bear only a remote relationship to the actual measurements. They are the result of processing by models and of multiple corrections[3,6].

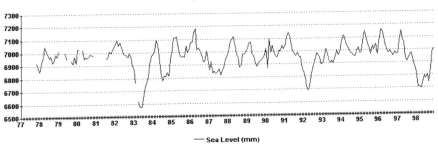

Figure 5.1   Mean Sea Level record for Funafuti, Tuvalu, from 1977 to 1999[4]

The researches of Mörner[7] of the International Association of Quaternary Research, which are critical of the IPCC methods, are largely ignored by them.

In Chapter 3 it is pointed out that many remote weather stations show no evidence of an overall temperature rise over the past century. Similarly, there are many of the more remote, and comparatively stable tide measuring stations which have not registered a rise. Also, many records showing a rise show a sudden jump rather than a smooth increase, suggesting a relationship with a particular event, such as the erection of tall buildings, or an airport, or a change of instruments.

As an example, many sea level records for the more remote Pacific islands show no sign of a rise over the past twenty years. Figure 5.1 for Funafuti, the capital of Tuvalu[4], shows that there has been no detectable change in sea level since 1978. Yet there are strident claims that Tuvalu is in danger of being swallowed up by the ocean. The New Zealand Government has agreed to special immigration concessions as a result of this false belief.

A recent National Geographic documentary called "Drowning Paradise"[8] devotes a whole hour to this forthcoming tragedy, for which there is no evidence.

Other Pacific islands showing no detectable change in sea level[4] are:

Tarawa, Kiribati, for 24 years
Nauru for 26 years
Honiara, Solomons, for 26 years,
Kanton Island for 28 years
Johnston Island for 50 years
Saipan for 22 years

Many others show a stable period followed by a sudden jump, most likely due to hotel or airport construction, or a hurricane. Most of them also show no mean temperature increase over the period. The El Niño events of 1983 and 1998 show low readings.

Since 1993 mean sea level has been measured by satellite altimeter observations. The latest record is shown in Figure 5.2[9].

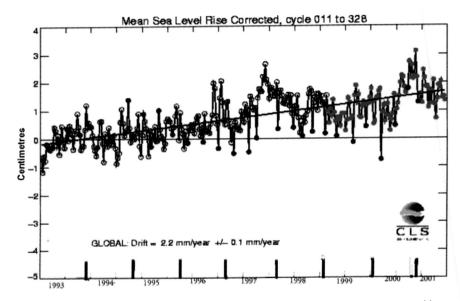

Figure 5.2   Mean sea level as measured by TOPEX/POSEIDON satellites since 1993[9]

On the face of it, it shows a rise in sea level over the period. Cabanes et al[10] have shown that the period from 1993 to 1998 is compatible with the ocean temperature measurements of Levitus et al[11], referred to in our Chapter 4. However, the later period is heavily influenced by the 1998 El Niño event; the satellites have had various problems of calibration and correction[12] and five years is an insufficient time to determine a climate trend. It is interesting that Cabanes et al find that the Topex/Poseidon measurements are double those estimated from tide gauges, direct evidence that the tide gauge information is from a distorted, biased sample.

All the calculations, estimates and forecasts of future sea level behaviour are dependent on the belief that the earth's surface temperature is increasing. If, as is argued in Chapter 3, this is not so, all these, together with the future projections of sea level rise given in Chapter 11 of *Climate Change 01*[13] can be regarded as completely without theoretical foundation.

To summarise: there is no firm evidence for recent rises in sea level unrelated to land movements and climate events such as El Niño.

References
1. *Climate Change 01* page 4
2. *Climate Change 01* page 639 "Changes in Sea Level"
3. *Climate Change 01* paragraph 11.3.2 Mean Sea Level Changes over the Past 100 and 200 years, pages 661–664
4. Permanent Service for Mean Sea Level (PSMSL) http://www.pol.ac.uk/psmsl/programmes
5. *Climate Change 01* page 661
6. Daly, J L. 2000. "Testing the Waters: A Report on Sea Levels", Greening Earth Society http://www.greeningearthsociety.org/Articles/2000/sea.htm
7. Mörner, N A. 1998. "Postglacial variations in the level of the sea: implications for climate dynamics and solid earth geophysics." *Review of Geophysics* 36 603–689
8. National Geographic Society Documentary 2001 "Drowning Paradise"
9. AVISO website http://www.jason.oceanobs.com/html/portail/actu/actu-welcome-uk.php3
10. Cabanes, C, A Casenave & C Le Provost. 2001. *Science* 294 840–842
11. Levitus, S, J I Antonov, T P Boyer & C Stephens. 2000. *Science* 287 2225–2229
12. *Climate Change 01* page 663
13. *Climate Change 01* Chapter 11, Figure 11

6. Computer Climate Models

D'Arcy Thompson in his "On Growth and Form" remarked "Numerical precision is the very soul of science" The main point of scientific laws is to be able to calculate from them.

When there is a project to send a man to the moon, the trajectory of the rocket has to be calculated using a complex set of mathematical equations with substituted parameters. The equations are based on scientific laws which have been established and tested, to known standards of accuracy. The parameters have also been measured to known standards of accuracy. In this way it is possible to predict exactly where the moon module will land, with a known measure of its accuracy.

The complex series of mathematical equations is a computer-based mathematical model. Before it can be used there must be statistically based studies on the accuracy of each part of the system, and comprehensive tests to prove that its predictions actually work within known limits.

Computer-based mathematical models have many applications. It is possible to simulate an entire industrial process and use the model to predict the effect of changes in the process.

The whole point is, that a computer-based mathematical model of any process or system is useless unless it has been *validated*. Validation of such a model involves the testing of each equation and the study of each parameter, to discover its statistically based accuracy using a range of numerically based probability distributions, standard deviations, correlation coefficients, and confidence limits. The final stage is a thorough test of the model's ability to predict the result of changes in the model parameters over the entire desired range.

No computer climate model has ever been validated. An early draft of *Climate Change 95* had a Chapter titled "Climate Models – Validation" As a response to my comment that no model has ever been validated, they changed the title to "Climate Models – Evaluation" and changed the word "validation" in the text to "evaluation" no less than fifty times. There is not even a procedure in any IPCC publication describing what might need to be done in order to validate a model.

Without a successful validation procedure, no model should be considered to be capable of providing a plausible prediction of future behaviour of the climate.

This same point, made more politely, so as to make its way through the hazards of peer review, is made in a recent paper by Soon et al.[1]

Instead of validation, and the traditional use of mathematical statistics, the models are "evaluated" purely from the opinion of those who have devised them. Such opinions are partisan and biased. They are also nothing more than guesses.

Attempts have been made to attach spurious measures of precision to these guesses. The following footnote appears on page 2 of the "Summary for Policymakers" of *Climate Change 01*[2]:

> *"In this Summary for Policymakers and in the Technical Summary, the following words have been used where appropriate to indicate judgmental estimates of confidence:* virtually certain *(greater than 99% chance that a result is true);* very likely *(90–99% chance):* likely *(66–90% chance);* medium likelihood *(33–66% chance);* unlikely *(10–33% chance);* very unlikely *(1–10% chance);* exceptionally unlikely *(less than 1% chance)*

As might be expected, there are no models or correlations falling into the *medium likelihood, unlikely* or very *unlikely* categories.

Chapter 8 of *Climate Change 01* "Model Evaluation"[3] evades the problem. A paragraph headed "What is meant by evaluation?"[4] never answers the question. They talk about "an approach" to evaluation. They confess *"We fully recognise that many of the evaluation statements we make contain a degree of subjective scientific perception and may contain much "community" or "personal" knowledge. For example, the very choice of model variables and model processes that are investigated are often based upon the subjective judgement and experience of the modelling community"*

In truth, all of their evaluation is subjective, and, since it is made by the modelling community itself, suspect.

The Executive Summary of the Chapter[5] contains numerous vague, subjective opinions.

- *"Coupled models can provide credible simulations"*
- *"Confidence in model projections is increased by the improved performance..."*
- *" There is no systematic difference..."*
- *" Some modelling studies suggest that..."*
- *"The performance of coupled models... has improved..."*
- *"Other phenomena previously not well simulated in coupled models are now handled reasonably well"*
- *"Analysis of, and confidence in, extreme events... is emerging"*
- *"Coupled models have evolved and improved significantly..."* (but we never get a numerical measure of "significant")
- *"Confidence in the ability of models to project future climates is increased by the ability of several models to reproduce the warming trend in the 20th century surface air temperature when driven by radiative forcing due to increasing greenhouse gases and sulphate aerosols"*

This statement illustrates the imperfect character of the IPCC "confidence".

Firstly, as explained in our Chapter 3, the warming trend of the combined weather station temperature measurements is most plausibly explained by their biased proximity to human habitation, and by such phenomena as volcanic eruptions, ocean and sun variability, none of which effects are incorporated in the models.

Secondly, model parameters, particularly those due to sulphate aerosols, are so uncertain, that it is possible to simulate almost any climate sequence, including a temperature fall, by suitable choices of parameters.

Thirdly a correlation, however successful, does not necessarily imply a cause and effect relationship.

Fourthly, the models are never applied to the more reliable temperature record in the lower troposphere, which shows no warming for the past 23 years.

Despite the entirely qualitative, inevitably prejudiced "assessments" of the IPCC, they conclude:

> "We consider coupled models, as a class, to be suitable tools to provide useful projections of future climates"[5]

All this, despite the fact that no model has ever provided a successful prediction of any future climate sequence.

The chapter continues with similar qualitative opinions which are too numerous to mention.

A number of spurious statistical procedures are used for "evaluation".

For example, it is common to provide a "range" of results, and consider this as somehow equivalent to an uncertainty figure. Of course this is nonsense. Each modellist will choose what he thinks are the best parameters and equations, but the "range" of results is not a fair measure of the probability distribution that would result if a proper statistical study were made.

An example is the treatment of "Climate Sensitivity", the predicted global mean temperature rise for a doubling of atmospheric carbon dioxide concentration, derived from many models. The "range" of results for the global temperature rise is quoted as between 1.5°C and 4.5°C. This figure was, apparently, originally derived by a "show of hands", a typically unscientific procedure. But this "range" does not begin to characterise the true uncertainties of model results.

Another dubious statistical procedure is "pattern analysis" where a pattern of climate data are compared with those predicted by a model. Invariably, no account is taken of the uncertainties of both components.

In a recent paper[6] Reilly et al. put it this way:

> "it is preferable to derive parameter uncertainty from observations, but the needed data often do not exist. Distributions of input parameters then must be selected by expert elicitation... Care must be taken in applying expert elicitation for well-known biases in human judgement"

Surely, when the "experts" are the modellists themselves the "well-known bias" immediately applies.

This article was followed by another by Allen et al[7] which concluded:

> "results are only of practical value when the factors responsible for the uncertainty are reasonably well documented and understood, which is certainly not the case for climate change in the late 21st century"

A further attempt to quantify guesswork is the use of "Bayesian Statistics"[8] where an initial guess is modified by subsequent supposedly better guesses.

Perhaps the best illustration of the huge uncertainties associated with all the climate models is Figure 6.1[9,10,11] which shows the global and annual mean radiative forcing for some of the model parameters

This diagram appears three times in *Climate Change 01*, each with a different caption. The caption in Chapter 6 is slightly more honest about the uncertainties.

First it should be noted that several of the most important contributors to radiative forcing are not even included. The most important greenhouse gas, water vapour, and the clouds that result from it, have been relegated to the status of a "feedback", where the large uncertainties in their estimation can be concealed.

The caption[11] says:

> "The forcing associated with stratospheric aerosols from volcanic eruptions is highly variable over the period and is not considered for this plot"

Then, indirect effects of tropospheric aerosols are left out, because they are "poorly understood", despite a statement in Chapter 5[12] that shows that they are well enough understood to say:

> "The largest estimates of negative forcing due to the warm-cloud indirect effect may approach or exceed the positive forcing due to long-lived greenhouse gases"

The caption to Figure 6.6[11] states:

> "The uncertainty range specified here has no statistical basis..."

Meaning that the uncertainties are larger than indicated. The use of the term "Level of Scientific Understanding" also implies much larger uncertainties.

The caption warns that an overall figure for radiative forcing cannot be obtained by merely adding and subtracting the figures in the diagram. But it is surely obvious that with the admitted uncertainties, a very large range of possible net radiative forcings are possible, including zero and negative values.

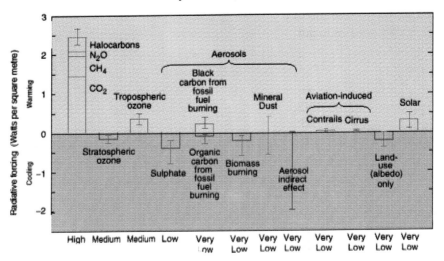

Figure 6.1   Global annual mean radiative forcings (in watts per square metre) for a number of agents since 1750[9,10,11]

Figure 6.1 surely shows that the uncertainties associated with the parameters commonly incorporated into climate models are so little known that the results of the models are completely worthless.

This conclusion is enhanced by reading Chapter 7, "Physical Climate Processes and Feedbacks"[12] of *Climate Change 01* which gives a detailed discussion of each of the processes, invariably concluding that the uncertainties are greater than is usually assumed by the models. But, of course, they decline to quantify any conclusion.

So far, the discussion has been mainly concerned with the general climate models; but exactly the same considerations apply to carbon cycle models[14]. We have some actual measured values for the carbon emitted by combustion of fossil fuels, and for the carbon in the atmosphere (if you accept that the Mauna Loa and other measurements can be considered representative). The other components of the cycle, are however, without known numerical value. There is a theoretical treatment of carbon dioxide absorption by the ocean, but no reliable measurements, apart from very rough "estimates" from isotope studies. The missing link is the carbon absorbed by the land surface, for which there are no reliable measurements, and also no reliable theory. Indeed, it used to be thought that there was a net outflow from the land, due to "deforestation.". The uncertainties in carbon cycle projections are thus without any numerical value for their uncertainties. Yet the IPCC has the effrontery to extrapolate "Stabilisation scenarios" as far ahead as the year 2300[15] with no indication of uncertainty. This is science fiction, not science.

Although models showed some success in predicting the temperature effects of the eruption of Mount Pinatubo in June 1991, the models have failed to successfully predict any other climate change. On the contrary:

- All the models predict that the Arctic and the Antarctic should warm much faster than the rest of the earth. This is just not happening.

- The models predict a temperature increase in the lower troposphere. Measurements for the last 23 years show that this is not happening.

- The models predict a steady increase in global surface temperature. The combined weather station record has changed in a fashion which is far from steady. Between 1940 and 1975 it showed a fall in temperature. Models can only cope with this by addition of arbitrary quantities of aerosols.

- Models predict that the Northern Hemisphere should warm at a slower rate than the Southern Hemisphere, because most aerosols are produced in the North. The combined weather station record shows greater warming in the North than in the South, and so does the satellite record in the lower troposphere.

- Models are unable to explain why most of the warming of the combined weather station record took place at night, or in the winter.

Despite the very great emphasis on models by the IPCC they have yet to show that their use, either to simulate climate, or to predict future climate, can be justified.

References

1. Soon, W, S Baliunas, S B Idso, K Y Kodratyev and E S Posmentier. 2001. "Modelling climatic effects of anthropogenic carbon dioxide emissions: unknowns and uncertainties." *Climate Research* 18 250–275
2. *Climate Change 01* "Summary for Policymakers", page 2, footnote 7
3. *Climate Change 01* Chapter 8 " Model Evaluation"
4. *Climate Change 01* Chapter 8, page 474
5. *Climate Change* Chapter 8 Executive Summary, page 473
6. Reilly, J, P H Stone, C E Forest, M D Webster, H D Jacobs, R G Prinn. 2001. "Uncertainty and Climate Change Assessment". *Science* 293 430–433
7. Allen, M, S Raper & J Mitchell. 2001. "Uncertainty in the IPCC's Third Assessment Report" *Science* 293 430–433
8. Barnett, T P, K Hasselmann, M Chelliah, T Delworth, G Hegerl, P Jones, E Rassmussen, E Roeckner, C Ropelewski, B Santer & S Tett.

1999. "Detection and Attribution of Recent Climate Change: A Status Report". *Bull Amer Meteorological Soc* 80 2631–2659
9   *Climate Change 01* "Summary for Policymakers", Figure 3, page 8
10 *Climate Change 01* "Technical Summary", Figure 9, page 37
11 *Climate Change 01* Chapter 6 "Radiative Forcing of Climate Change", Figure 6.6, page 392
12 *Climate Change 01* Chapter 5 "Aerosols, Their Direct and Indirect Effects", page 334
13 *Climate Change 01* Chapter 7 "Physical Climate Processes and Feedbacks", 417–70
14 *Climate Change 01* Chapter 3 "The Carbon Cycle and Atmospheric Carbon Dioxide". 3.6. Carbon Cycle Model Evaluation, pages 213–218
15 *Climate Change 01* Chapter 3, page 223

7. Forecasting The Future

Forecasting the future can hardly be considered as a science because there are too many unknown variables. Weather forecasters are expected to tell us what the weather will be, but despite a large array of scientific equipment they are still often wrong. Economists are employed to predict the future behaviour of the economy, but they are rarely right. It has been said that they are the only professionals who continue to be employed even when they are always mistaken.

Because of future uncertainties, economists are usually reluctant to provide forecasts further ahead than a decade or so. Yet climate scientists are now routinely telling us what will undoubtedly happen a hundred years, or sometimes several hundred years ahead.

The previous Chapter has shown that computer models of the climate, based on the assumption that the greenhouse effect is the only influence on the climate, have never successfully predicted any future climate change, and have such large uncertainties that they are quite unsuitable for forecasting the future.

The models usually calculate the consequences of increasing atmospheric carbon dioxide concentrations, typically the effect of it doubling. Such a model does not, however constitute a future prediction unless it also includes a forecast of how long it will take for the concentration of carbon dioxide to double.

In order to do this, it must have a *scenario* of how carbon dioxide and other greenhouse gases will change their atmospheric concentrations in the future. Then, there has to be a model which converts greenhouse gas emissions to atmospheric concentrations, and then the climate model which can convert these concentrations to *radiative forcing*, the extra radiation at the top of the atmosphere due to the greenhouse gas changes. You then require yet another model to convert this forcing to a projected temperature or sea level change at the earth's surface.

It should be made plain that all the estimates provided in the scenarios are purely a matter of the opinions of "experts", comparable to the opinions of the "confidence" to be placed on models.

A range of *scenarios* has been used by the IPCC in order to forecast, predict, or as they would claim to prefer it, *project* the future.

The typical, and still a highly popular scenario, is the assumption that the carbon dioxide concentration in the atmosphere will increase by 1% a year. This assumption is made by 30 of the 99 models listed in Chapter 9 of *Climate Change 01*[1].

The assumption merely multiplies the measured rate of rise of CO_2 of 0.4% per year for the past 27 years, by a factor of 2½, and is typical of the wild, irresponsible exaggeration which pervades almost all of the IPCC scenarios.

Climate Change 01[2] gives the following excuse.

> "A common standardised forcing scenario specifies atmospheric CO_2 to increase at a rate of 1% a year compound until the concentration doubles (or quadruples) and is then held constant. The CO_2 content of the atmosphere has not, and likely will not, increase at this rate (let alone suddenly remain constant at twice or four times an initial value). If regarded as a proxy for all greenhouse gases, however, an "equivalent CO_2" increase of 1% a year does give a forcing within the range of the SRES [Special Report on Emission Scenarios] scenarios"

In addition to admitting the gross exaggeration involved in multiplying a well-established rate of rise in CO_2 by 2½, this passage then claims that there could be a 1% rise in "equivalent CO_2", which includes other greenhouse gases. Yet the most important of the other greenhouse gases, methane, has been increasing at a declining rate for the past 14 years, and its concentration is currently falling. The only way the scenarios can provide a 1% increase in "equivalent CO_2" is to combine an exaggerated rise in CO_2 itself with a reversal of the recent methane trend, and a forecast concentration increase. The above passage shows that the irresponsible exaggeration of 1% CO_2 increase a year still pervades the most recent, SRES (Special Report on Emission Scenarios) scenarios.

There have been three sets of more detailed scenarios than the 1% solution.

The first, launched with *Climate Change 90*[3] consisted of four scenarios, A, B, C, and D. Scenario A was termed "Business As Usual" (BaU) or as SA90, and was considered by many as a plausible forecast for the future.

The next series of six scenarios was developed for *Climate Change 92*[4], and the details are available from a supplementary Report[5]. The scenarios were designated IS92a, IS92b, IS92c, IS92d, IS92e and IS92f.

A new set of scenarios was used in *Climate Change 01*[6] The scenarios were prepared by a special committee of IPCC Working Group III. Their Report, *Special Report on Emission Scenarios*[7], was produced without any input from the scientists involved with Working Group I, and its conclusions were foisted on *Climate Change 01* without opportunity for discussion. Six teams of specialists from 18 countries drew up a total of 40 scenarios, all based on four "storylines" claimed to represent different views of what might happen in the future. The 40 scenarios were originally summarised in the form of four "Marker Scenarios", A1, A2, B1. and B2, but as time went on, they split A1 into three; A1F1, A1B and A1T.

The existence of all these different scenarios has led to considerable confusion, as the modellists did not know which to use. *The Special Report on Emission Scenarios 00*[7] was published a year before *Climate Change 01*, which has provided additional information on the scenarios which was not present in the *Special Report*. Additional confusion was caused by partial attempts to correct the IS92 scenarios to remedy the fact that they were unable to successfully predict even 10 years ahead[8]. Only one of these has been currently listed[9], an amended version of the carbon dioxide and methane concentrations for IS92a, which, confusingly, has the same name, while the original IS92a is referred to as IS92a/SAR. Presumably there are also amended versions of the other IS92 scenarios which are not revealed.

The IPCC has gone to some trouble to explain that the scenarios should not be used to forecast the future.

Climate Change 92[10], says:

> "Scenarios are not predictions of the future and should not be used as such."

The *Special Report on Emissions Scenarios 00*[11] puts it this way:

> "Scenarios are images of the future or alternative futures. They are neither predictions nor forecasts."

Despite these assertions the scenarios are widely presented as predictions and forecasts, not only by the media, politicians and "activists" of various colours, but often by the scientists who are responsible for them, in their public appearances. The "projections" are also assumed to be genuine forecasts by *Climate Change 01: Impacts*[12] and by *Climate Change 01: Mitigation*[13].

The *Special Report on Emissions Scenarios 00*[14] also insists:

> "The possibility that any single emissions path will occur as described in the scenario is highly uncertain", and

> "No judgement is offered in this Report as to the preference for any of the scenarios and they are not assigned probabilities of occurrence, neither must they be interpreted as policy recommendations"

This statement amounts to an endorsement of the wildest unlikely scenarios. The scenarios, are, as a matter of course, used extensively as policy recommendations despite this caveat.

Also[13]:

> "there is no objective way to assign likelihood to any of the scenarios. Hence there is no "best guess" or "business-as-usual" scenario"

There is an obvious objective way of assigning likelihood which is ignored by the IPCC. Projections that prove to be correct over a period of, say, ten years, are more likely to be reliable than projections that are seriously incorrect over the same period.

The most priceless statement is[15]:

> *"These tools are less suitable for analysis of near-term developments and this report does not intend to provide reliable projections for the near term"*

They simply do not care that their "projections" differ from current or "near-term" reality, and they emphasise the "long term" because they will never have to answer for their incompetence until all of us are dead and gone. Near-term projections that are unreliable must surely cast serious doubt on the credibility of long-term projections, particularly, as will emerge, when most of them are seriously exaggerated.

Climate Change 94[16] says:

> *"Scenarios deal with the future so they cannot be compared with observations"*

Unfortunately, the future tends ultimately to become the present, and then the past. It then becomes possible to compare the scenarios with what has actually happened, and to assess whether they have been successful.

Tables 7.1 to 7.5 show the assumptions made by the various scenarios between the years 1990 and 2100 for a range of parameters. They are compared with the actual measured quantities for the years 1990 and 2000. Scenario figures are from *Climate Change 90* for the A, B. C and D scenarios, Pepper et al[5] for the IS92 scenarios, and *Special Report on Emissions Scenarios 00*[7] and *Climate Change 01*[9] for the SRES scenarios. Figures often had to be interpolated by estimates from enlarged graphs.

Population

Table 7.1 shows how scenario estimates compare with measured population figures. The measured data are from the UNFP[17], and the US Census Bureau[18].

All the "projected" values for the year 2000 are exaggerated. by amounts between 0.6% and 6% Of the IS92 scenarios the most successful were IS92c and IS92d. The IPCC continues to place its faith on IS92a, however.

TABLE 7.1 Scenario estimates of world population (in billions)

Scenario	1990	2000	2010	2020	2100
Measured, UNFP	5.29	6.06			
Measured, US Census	5.28	6.07			
SA90 (BaU)	5.25	6.21	7.06	7.50	10.36
IS92a	5.25	6.21	7.11	7.78	11.31
IS92b	5.25	6.21	7.11	7.78	11.3
IS92c	5.25	6.10	6.61	7.22	6.42
IS92d	5.25	6.10	6.61	7.22	6.42
IS92e	5.25	6.21	7.11	8.69	11.31
IS92f	5.25	6.42	7.56	8.69	17.59
A1F1	5.29	6.12	6.89	7.6	7.14
A1B	5.29	6.12	6.89	7.5	7.14
A1T	5.26	6.12	6.89	7.6	6.98
A2	5.29	6.17	7.18	8.2	15.1
B1	5.29	6.11	6.92	7.6	6.98
B2	5.29	6.12	6.92	7.6	10.41

The SRES scenario assumptions are too high for the year 2000 despite the publication date of 2001, with A2 nearly 2% too high

Coal Production

An example of the future projections for energy usage in the scenarios can be shown by the figures for coal production, Table 7.2.

Figures are from the United Nations Energy Handbook[19], updated by BP.[20]

IS92 scenarios all exaggerate coal production in the year 2000, by between 3 and 17%. As before, the most successful scenarios are IS92c and IS92d. IS92e overestimates the 2000 figure by 17%, assumes an increase of 65% by 2010 and envisages an increase in world coal production of eleven times over the 2000 figure by the year 2100.

The 1990 figures for the SRES "illustrative" scenarios are something of a mystery. They are obtained by multiplying "Share of coal in primary energy" by "primary energy" in Table 2a of the "Summary for Policymakers of *Special Report on Emissions Scenarios 00*[21]. Individual scenarios[22] give much higher figures for 1990 that vary between 82 and 105 EJ. This, surely, shows the sloppiness that pervades these scenarios. There was, after all, a measured value of coal production of 96EJ in 1990.[19]

TABLE 7.2 Scenario Estimates of Coal Production (in Exajoules)

Scenario	1990	2000	2010	2020	2100
Measured	96	103			
IS92a	99	117	158	211	680
IS92b	99	114	154	201	650
IS92c	99	106	125	144	127
IS92d	99	110	145	179	333
IS92e	99	115	170	258	1126
IS92f	99	120	164	236	871
A1F1	84	115	150	194	601
A1B	84	111	135	165	89
A1T	84	110	130	149	20
A2	84	95	110	131	910
B1	84	95	115	133	41
B2	84	88	91	96	146

Despite the initial low figure, the A1 scenarios exceed the coal production figure for 2000 by as much as 16%.

A1F1 envisages a coal industry in 2100 7.2 times the size of 1990, and A2 10.8 times. The scenario A1C AIM[22] puts it up to 12.2 times. On the other hand, A1T projects a coal industry one quarter the size of 1990 by 2100, and the scenario B1T MESSAGE[22] gets it down to 2%. These extreme values are rather absurd.

Carbon Dioxide Emissions

Table 7.3 shows the assumed emissions of carbon dioxide from fossil fuels and cement, compared with the measured values[23,20]. Figures assuming that the current rate of increase since 1974 (0.8 GtC/yr) might continue have been included. The *Special Report on Emissions Scenarios*[21,22] confuses us by adding (or subtracting) emissions from land use changes to give "total anthropogenic emissions", but the figures for fossil fuel combustion have been chosen here.

Most of the figures appear to include emissions from cement production. The *Special Report on Emissions Scenarios 00*[21,22] includes "industrial processes".

TABLE 7.3 Emissions of CO_2 by fossil fuel combustion, in GtC/yr as projected by the IPCC

Scenario	1990	2000	2010	2020	2100
Measured	6.1	6.8			
0.08GtC/yr	6.0	6.8	7.2	8.0	14.8
SA90	6.7	8.1	9.0	10.6	22.7
IS92a	6.1	7.1	8.68	10.26	28.91
IS92b	6.1	7.0	8.5	10.0	19.2
IS92c	6.1	6.2	6.8	7.4	4.9
IS92d	6.1	6.6	7.5	8.5	10.5
IS92e	6.1	7.8	10.3	12.7	36.0
IS92f	6.1	7.5	9.7	11.9	26.6
A1F1	6.0	6.9	8.65	11.19	30.32
A1B	6.0	6.9	9.68	12.12	13.10
A1T	6.0	6.90	8.33	10.00	4.31
A2	6.0	6.90	8.46	11.01	28.91
B1	6.0	6.90	8.50	10.00	5.20
B2	6.0	6.90	7.99	9.02	13.82

The IS92 scenarios had one scenario (IS92c) which seriously assumed that there would be hardly any increase in emissions from 1990 to 2000, and another (IS92d) which assumed an increase of only 0.2GtC less than what actually happened. Apart from these two, however, all the rest of the IS92 scenarios grossly overestimated the actual emissions figure for the year 2000 by between 3 and 5%. The SA90, ("Business as Usual") scenario, overestimated 2000 by 19%.

All of the SRES scenarios expect a sudden increase. as of now, in the rate of emission shown by the last 26 years' trend (0.8% a year), to give figures for the year 2010 between 11 and 34% above that recent long-term trend. A1F1 and A2 more than double that trend by 2100. However, A1B and B2 have got back to the current trend by 2100, and A1T and B1 are below the 1990 emissions by 2100.

Carbon Dioxide Concentrations

Table 7.4 shows the atmospheric carbon dioxide concentrations used in the various scenarios, compared with the measured values[24], which have shown a constant rate of rise of 0.4% a year since 1974. The effects of this rate continuing are shown.

An innovation, in *Climate Change 01*[9], is that the carbon dioxide concentrations resulting from the use of different carbon cycle models are given, so that there is a range of figures, depending on the model chosen. This example should be followed more often, as it reminds us how uncertain these figures really are. In Table 7.4 are included the results from the highest (TOP) and the lowest (BOTTOM) individual scenarios.

The differences between the different models illustrates the fact that the model outputs have very large uncertainties, which are much higher than the differences between individual scenarios.

Of the IS92 scenarios IS92c and IS92d are compatible with a continuing 0.4% a year. IS92e is the only one following the 1% solution, but IS92a gets there if you use the right model.

A1F1, and A2 have carbon dioxide concentrations for 2100 which are in the region of the 1% increase a year, again, with the right model. The TOP assumption, is well above even this level.

TABLE 7.4 Atmospheric Carbon Dioxide Concentrations in ppmv assumed by the IPCC scenarios

Scenario	1990	2000	2010	2020	2100
Measured	354	369			
0.4% a year	354	369	384	400	550
1% a year	354	391	432	477	1057
SA90	354	373	400	425	827
IS92a SAR	353–354	370–372	391–393	416–418	709–715
IS92a TAR	352–353	367–369	382–396	401–431	640–902
IS92b	354	374	393	415	659
IS92c	354	373	386	400	479
IS92d	354	373	386	400	530
IS92e	354	375	400	423	906
IS92f	354	375	400	420	707
A1F1	352–353	367–369	381–394	405–431	824–1248
A1B	352–353	367–369	383–397	405–431	617–918
A1T	352–353	367–369	381–394	398–422	506–735
A2	352–353	367–369	382–395	402–427	735–1080
B1	352–353	367–369	380–394	398–422	486–681
B2	352–353	367–369	380–398	394–417	544–769
TOP	354	369	404	436	1256
BOTTOM	354	364	370	394	485

Methane Concentrations

TABLE 7.5 Methane concentrations assumed by the IPCC scenarios in ppbv

Scenario	1990	2000	2010	2020	2100
Measured	1693	1753			
Current Trend	1693	1753	1745	1737	1671
SA90 (BaU)	1759	1972	2237	2484	4018
IS92a	1700	1810	1964	2145	3616
IS92a TAR	1700	1760	1855	1979	3136
IS92b	1700	1810	1964	2145	3616
IS92c	1700	1787	1880	1984	2069
IS92d	1700	1787	1878	1975	2146
IS92e	1700	1824	2007	2224	4291
IS92f	1700	1817	1995	2221	4669
A1Fi	1700	1760	1851	1986	3413
A1B	1700	1760	1871	2026	1974
A1T	1700	1760	1856	1998	2169
A2	1700	1760	1861	1997	3731
B1	1700	1760	1827	1891	1574
B2	1700	1760	1839	1936	2973

Table 7.5 shows the atmospheric methane concentrations assumed by the IPCC scenarios, compared with the measured values[24]. As noted previously, the rate of increase of methane in the atmosphere has been falling for the past 17 years, and the actual concentration is currently falling. This trend must surely continue for a while, at least, and it may turn out to be a long-term effect. None of the scenarios are prepared to recognise this, even as a possibility.

This whole Table therefore has an air of unreality: a total unwillingness to believe actual measured values, or their trends. Only scenario B1 has a methane concentration in 2100 corresponding to the trend of the past 16 years, but even this scenario goes well above the trend as of now, so it assumes an immediate sudden reversal of the current trend. The whole set of scenarios can be dismissed as completely unrealistic from this Table alone.

Temperature

The whole purpose of the IPCC is to provide estimates of the global temperature rise in the future, supposedly as a result of increases in greenhouse gases. In order to do this two sets of unreliable, unvalidated models (both climate models and carbon cycle models) are applied to mostly highly unlikely emissions scenarios, some of whose assumptions are described above. This means that almost any result can be obtained, depending on the choice of model, model parameters and scenario. This choice, therefore can readily be manipulated to secure a desired result, which is a set of projections which satisfies the political authorities that sponsor the research.

The process is well illustrated by the history of the currently published set of IPCC temperature projections.

Climate Change 90[25] combines a range of "climate sensitivity" (High estimate 4.5°C, Best Estimate 2.5°C and Low Estimate 1.5°C) with the four scenarios A, B, C and D to give Figure 8, page xxii, which shows simulated temperature rise between 1765 and 2100 for the three Climate Sensitivities and Scenario A (Business as Usual, BaU). It shows a rise over this period of between 2.9°C and 6.2°C (1990 to 2100, 2.1°C to 4.8°C).

Figure 9, page xxiii, shows the rise from 1765 and 2100 for the Best Estimate climate sensitivity, and the four scenarios; a rise between 2.0°C and 4.2 °C (1990–2100 of 1.0°C to 3.2°C).

The Executive Summary[26] states:

> "This will result in a likely increase in global mean temperature of about 1°C above the present value by 2025 and 3°C before the end of the next century". They then point out that the lowest scenario, D would give about 0.1°C per decade

Climate Change 95, on pages 322 and 323[27] has five figures giving possible temperature changes between 1990 and 2100.

Figure 6.20 (page 322) uses scenario IS92a, four different climate sensitivities and changing and constant aerosols. The range of temperature increase from 1990 to 2100 is from 1.4°C to 3.5°C.

Figure 6.21 (page 322) uses IS92a and climate sensitivity 2.5°C, CO_2 forcing alone, and with aerosols. The temperature rise 1990 to 2100 ranges from 1.8°C to 2.4°C.

Figure 6.22 (page 323) shows the effect of all the IS92 emissions scenarios using full aerosol forcing and climate sensitivity of 2.5°C. The temperature increase 1990 to 2100 ranges from 1.3°C to 2.5°C.

Figure 6.23 shows the effect of IS92 scenarios and constant aerosol, with climate sensitivity 2.5°C. The range of increase 1990 to 2100 is 1.3°C to 3.2°C.

Figure 6.24 shows the extreme range of possible changes, between IS92c with constant aerosol and climate sensitivity of 1.5°C, giving a rise of 0.8°C, to IS92e with constant aerosol and climate sensitivity of 4.5°C, giving a rise of 4.5°C.

All these Figures used a single carbon cycle model.

It is difficult from this variety to decide which figures to take, and the "Summary for Policymakers" expresses this difficulty, as follows[28]:

"For the mid-range IPCC emission scenario IS92a, assuming the "best estimate" value of climate sensitivity, and including the effects of future increases in aerosols, models project an increase in global mean surface air temperature relative to 1990 of about 2.0°C by 2100. This estimate is approximately one third lower than the "best estimate" in 1990... Combining the lowest IPCC emission scenario with a "low" value of climate sensitivity and including the effects of future changes in aerosol concentrations leads to a projected increase of about 1°C by 2100. The corresponding projection for the highest IPCC scenario (IS92e) combined with a "high" value of climate sensitivity gives a warming of about 3.5°C".

There was little exaggerated public attention drawn to the obviously extreme estimate of 3.5°C warming by 2100 after the report was published.

Figure 7.1 shows the current IPCC projected range for the temperature rise from 1990 to 2100 from Figure 9.14 of Chapter 9 of *Climate Change 01* which is the same as Figure 22a of the Technical Summary[29] A slightly different Figure appears as Figure 5(d) of the *Summary for Policymakers*[30]. For the first two the alleged range between IS92c low and IS92e high is from 1.2°C to 3.4°C. For the

Summary for Policymakers diagram[28] the range shown for "All IS92" is between 1.0°C and 3.5°C.

The big change from *Climate Change 95* is the upper end of the range of the increase in projected temperature rise from 1990 to 2100, which is now 5.8°C instead of 3.5°C. There is also an increase in the lower end of the range from 1.0°C to 1.4°C.

Climate Change 01[31] explains this as follows :

> *"It is evident that the range in forcing for the new SRES scenarios is wider and higher than in the IS92 scenarios. The range is wider due to more variation in emissions of non-CO_2 greenhouse gases".*

The range for the annual emissions of the most important non-CO_2 greenhouse gas, methane, for the year 2100, changed only from 546-1072Mt for the IS92 scenarios to 236-1069Mt, which should have led to a reduction in the lower range figure.

Then[31]:

> *"The shift to higher forcing is mainly due to the reduced future sulphur dioxide emissions of the SRES scenarios compared to the IS92 scenarios"*

It is true that the range of sulphur emissions of 87-254 MtS/yr for the IS92 scenarios for the year 2100 was reduced to 11-83 MtS/yr for the SRES scenarios, but it was not explained why increased amounts of use of coal with increasing sulphur content could justify such a reduction.

But this is not the whole story, as can be seen by studying the First and Second Drafts of *Climate Change 01*.

In the First Draft Figure 9.13a on page 66 showed a range of projected temperature increase between 1990 and 2100 of between 1.7°C and 4.0°C.

In the Second Draft, Figure 9.18, page 88 showed a range of projected temperature increase from 1990 to 2100 of between 1.3°C and 5.0°C.

Since the effects of lower sulphur dioxide had already been incorporated into the First Draft, the escalation from 4.0°C to 5.8°C in the maximum of the range must have been secured by other means.

This was done by two devices. Firstly the original four "Marker" Scenarios from the 40 SRES scenarios were expanded into six, notably by scenario A1F1 which has an extremely high use of fossil fuels. Secondly, if that was not enough, the range was increased still further by applying the "envelope of all scenarios" to Figure 7.1, so that the maximum assumptions of all the scenarios could be utilised.

Then, instead of using only one climate model, as was the previous custom, no less than seven models were used, particularly a very high one, GFDL-R15-a.

Section 9.3.3 of *Climate Change 01*[32] is a record of how Figure 7.1 was designed to suit the requirements of the politicians, by including some uncertainties, omitting others, choosing parameters to suit, and leaving others out. It is perfectly clear that by a suitable choice of parameters, models and uncertainties any required temperature projection can be supplied. They make this point clear in the following passage:

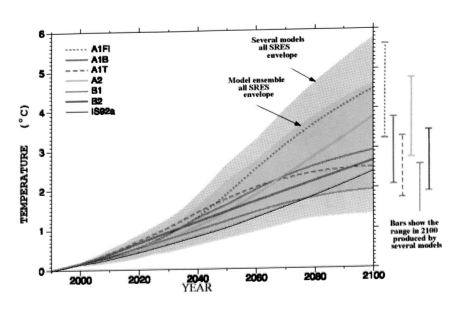

Figure 7.1   Current IPCC temperature projections[29]

"*The range for these two parameter settings for the full set of SRES scenarios is 1.4 to 5.8°C. Note that this is not the extreme range of possibilities, for two reasons. First, forcing uncertainties have not been considered. Second, some AOGCMs have effective climate sensitivities outside the range considered.*"

They say this in order to suggest that the upper end of the range could be even higher, but they fail to mention that the lower end of the range could also extend downwards, to a zero or negative temperature change. If they were really honest they would confess the truth, which is that the models and the projections are compatible with almost any conceivable temperature change in the future and are therefore worthless as a guide to future action.

References

1. *Climate Change 01* Chapter 9, "Projections of Future Climate Change". Table 9.1, pages 538–540
2. *Climate Change 01* Chapter 9 page 533
3. *Climate Change 90* Appendix 1 page 341
4. *Climate Change 92* Chapter A3 "Emissions Scenarios for IPCC: An Update", pages 69–95
5. Pepper, W, et al. 1992. "Emission Scenarios for the IPCC: An Update." IPCC supplementary paper
6. *Climate Change 01* page 799
7. *Special Report on Emissions Scenarios 00*
8. Gray, V R. 1998. "The IPCC future projections: are they plausible?" *Climate Research* 10 155–162
9. *Climate Change 01* Appendix II, SRES Tables pages 807–810
10. *Climate Change 92* page 9
11. *Special Report on Emissions Scenarios 00* Technical Summary, page 1
12. *Climate Change 01: Impacts*
13. *Climate Change 01: Mitigation*
14. *Special Report on Emissions Scenarios 00*. Summary for Policymakers page 3
15. *Special Report on Emissions Scenarios 00*. Technical Summary page 2
16. *Climate Change 94* page 252
17. United Nations Population Fund. Annual Reports

18. United States Census Bureau http://www.census.gov
19. *United Nations Energy Handbook*, current year.
20. British Petroleum Energy Statistics http://www.bp.com
21. *Special Report on Emissions Scenarios 00.* Summary for Policymakers, Table 2a page 91
22. *Special Report on Emissions Scenarios.* Appendix VII, Tables
23. Trends; Carbon Dioxide Information and Advisory Center. http:/cdiac.esd.ornl.gov/trends
24. Carbon Cycle Group, NOAA, http://www.cmdl.noaa.gov/ccg
25. *Climate Change 90* pages xxii and xxiii
26. *Climate Change 90* page xi
27. *Climate Change 95* pages 322 and 323
28. *Climate Change 95* page 5
29. *Climate Change 01* Chapter 9, page 555, "Technical Summary" page 70
30. *Climate Change 01* "Summary for Policymakers", page 14
31. *Climate Change 01* page 555
32. *Climate Change 01* pages 554–557

8. Extreme Events

The earth's climate has undergone enormous changes since the earth cooled to a temperature capable of supporting life, some 3 billion years ago. Details of past changes are often difficult to characterise or measure, even after painstaking study of geological and ocean deposits. There have certainly been many changes in the past which greatly exceed human experience.

It is quite possible that changes in the concentration of even such a relatively unimportant greenhouse gas as carbon dioxide might have an influence on the climate, apart from its established influence on plant growth. The belief in some sort of influence has led to efforts to dredge the world press for examples of "unusual" weather events, which are frequently attributed to carbon dioxide changes despite the ancient principle that a correlation, however convincing, does not establish a cause and effect relationship.

While it is possible to measure such things as gas concentrations or temperature, there is no satisfactory way of measuring the frequency of past hurricanes, floods, droughts, or snow or ice coverage. How much wind constitutes a hurricane? How dry is a drought? Where is the snow or ice boundary? It is therefore not possible to provide a fair comparison of current experience with the extent or importance of these calamities over the years. The damage or inconvenience of these events is a function of the size and complexity of the neighbouring human population. Insurance pay-outs must inevitably rise as there are more people with higher-valued property.

The IPCC[1] has expressed their level of confidence in the possible future increase of some of these unusual weather patterns by their *estimated* (for which read *guestimated*) likelihood levels, "based on expert judgement", but the data are shoddy, and the models used are, as usual, untested Many of these events are anecdotal rather than representative, and negative events are preferred to positive ones. There are even attempts to argue that better weather is harmful. The whole of *Climate Change 01: Impacts*[2] is devoted to the proposition that almost everything is going to get worse. The basic assumption throughout, that future warming is inevitable, is behind most of the projections, and is, as has been pointed out in our Chapter 3, by no means an established fact. Lomborg[3] has shown what happens to the pessimistic forecasts when reliable statistics are scrutinised.

References
1. *Climate Change 01* "Summary for Policymakers" page 15
2. *Climate Change 01: Impacts*
3. Lomborg, B. 2001. *The Skeptical Environmentalist.* Cambridge University Press

9. CONCLUSIONS

IPCC Conclusions

Climate Change 90 in its first paragraph, began with the fundamental tenet of the greenhouse theory:

> "*emissions resulting from human activities are substantially increasing the atmospheric concentrations of the greenhouse gases... These increases will enhance the greenhouse effect, resulting on average in an additional warming of the Earth's surface.*"[1]

They did not admit that this additional warming was measurable, significant, or harmful. As evidence for their statement they placed exclusive emphasis on the temperature increase shown by the combined surface measurements. They were careful not to attribute this warming to increases in greenhouse gases, or even to human influence when they stated:

> "*The size of this warming is broadly consistent with the predictions of climate models, but is also of the same magnitude as natural climate variability.*"[2]

The claim that the temperature change is "broadly consistent with the predictions of climate models" is disputable. The only "warming" considered was that of the combined surface temperature record of weather stations and ships, which are not distributed uniformly over the earth's surface, and are biased by their predominant proximity to cities, buildings and other human activity. The temperature records of the troposphere, from balloons and satellites, were ignored, despite the statement:

> "*the upper troposphere shows that there has been a rather steady decline in temperature since the late 1950s and early 1960s, in general* disagreement *with model simulations*

that show warming at these levels when the concentration of greenhouse gases is increased."[3]

Then, the models predicted a uniform temperature rise, when the combined surface record was highly non-uniform, even showing a fall in temperature from 1940 to 1975, quite incompatible with the models.

Finally the models available at the time predicted a much greater temperature rise than that shown by the surface record.

It would be more accurate to say that the models were "broadly inconsistent" with the predictions of climate models.

Climate Change 95 concluded:

"The balance of the evidence suggests a discernible human influence on global climate."[4]

This statement was deliberately vague and ambiguous. It did not claim that there *is* a discernible human influence on the climate, merely that the "balance of the evidence" "suggests" it.

The statement also failed to identify the "human influence" which is supposedly "discernible".

It is obvious to most of us that humans influence the climate in a variety of ways. Building cities, planting or harvesting forests, and discharging atmospheric pollution all influence the climate. It is impossible to disagree with the above statement. The only one of these influences considered by the IPCC was "deforestation".

The press and politicians have interpreted the statement to imply that the "discernible human influence" is an increase in greenhouse gas emissions, and the IPCC have never tried to discourage this false interpretation of their conclusion.

The false interpretation was encouraged because *Climate Change 95* devoted much attention to climate models which assume greenhouse gases as the only influence on the climate, but very little attention to possible other "human activities" which may influence it.

Climate Change 01 reasserted the above conclusion from *Climate Change 95* and also concluded:

"There is new and stronger evidence that most of the warming observed over the last 50 years is attributable to human activities."[5]

CONCLUSIONS

and

> "*in the light of the new evidence and taking into account the remaining uncertainties, most of the observed warming over the last 50 years is likely to have been due to the increase in greenhouse gas concentrations.*"[5]

By selecting "the last 50 years" this statement deliberately ignored the temperature records from weather balloons (for the last 41 years) and satellites (for the past 21 years) which did not show evidence of significant warming. They were therefore selecting the combined weather station and ship record, with all its obvious bias, as their only evidence of "observed warming".

Since that record shows an "observed cooling" between 1950 an 1975, half of the selected range, the statement makes little sense. What they are really saying is that the "observed warming" in the combined surface record has only been evident for the past 25 years, without mentioning that it disagrees sharply with the observed temperature record from weather balloons and satellites over the same period.

This time they specified the "human activity" that is supposed to be responsible for the "warming" of the combined surface record, as "likely" to be due to an increase in greenhouse gas concentrations. "Likely" is subjectively estimated as implying a 66-90% chance of being correct. There was nothing in the body of the Report to support this estimated high probability. A fall in surface temperature from 1950 to 1975 surely proved it to be wrong.

Despite this apparently confident attribution, Chapter 1 of *Climate Change 2001* contradicted it, as follows:

> "*The fact that the global mean temperature has increased since the late 19th century and that other trends have been observed does not necessarily mean that an anthropogenic effect on the climate has been identified. Climate has always varied on all time-scales, so the observed change may be natural.*[6]

This is where we came in.

Our Conclusions

This book agrees with the IPCC that the rise in temperature shown by the combined weather station/ship record for the past 25 years is attributable to human activities. But we disagree with their statement that it was likely to be caused by an increase in greenhouse gases.. The most likely "human activity" responsible is the obvious bias in favour of cities, the increase in extent of cities, roads and airports, the increased energy consumption, the increase in vehicles and aircraft, and the growth of vegetation around many weather stations.

No convincing evidence has been presented by the IPCC, or anyone else, that a surface temperature increase has resulted from increases of greenhouse gases.

Our conclusion that greenhouse gases are not responsible for the surface temperature rise is supported by the tropospheric temperature measurements, which do not show a significant rise, when this region is supposedly that most influenced by greenhouse gases. Temperature measurements in remote locations, corrected figures for the USA, and most proxy measurements support the conclusion.

It is agreed that increases in greenhouse gases have taken place and that they must have an influence on the climate. However, there is no evidence that this influence has been identified or measured. There is certainly no evidence that it is harmful, and much evidence that it is beneficial.

We reject computer climate models in their entirety. They assume that the enhanced greenhouse effect is the only influence on the climate, in the teeth of the obvious existence of many other factors.

The parameters and equations used in the models are very uncertain and there are no truly quantitative estimates of the accuracy of model predictions. We reject "expert opinion" as a scientifically acceptable measure of model reliability. Models can be "adjusted" to fit some climate sequences because of the wide range of choice of parameters and equations. Besides being readily fitted to the combined surface temperature record, they can also be "adjusted" to be perfectly compatible with a stable climate or a cooling climate, merely by a different choice of parameters. Because of these uncertainties they have therefore little value for climate prediction.

No model has ever been *validated* in a scientifically acceptable manner. Successful prediction of a wide range of future climate changes is an essential prerequisite for the usefulness of any model to be used for future prediction.

Future forecasts presented by the IPCC are nothing but informed, but heavily biased guesses, processed by untested models. The forecasts are therefore easily manipulated to comply with current political expectations or demands. They have been shown to be incapable of successfully prediction of climate or social changes for even ten years ahead and should therefore be rejected as reliable guides to future climate for as long as a hundred years ahead.

There is no acceptable scientific basis for the reductions of greenhouse gas emissions called for by the Framework Convention on Climate Change or for the Kyoto Protocol. These instruments would merely apply economic burdens with no measurable result on the climate.

References

1   *Climate Change 90*, page xi
2   *Climate Change 90*, page xii
3   *Climate Change 90*, page 221
4   *Climate Change 95*, page 4
5   *Climate Change 01*, page 10
6   *Climate Change 01*, page 97

A Note on Sources

Most of the references in this book are from the recently published Report of Working Group 1 of the Intergovernmental Panel on Climate Change, *Climate Change 2001: The Scientific Basis*, Some are from previous Reports from this committee, and some from the reports of Working Groups II and III. These Reports contain a comprehensive review of the scientific and economics literature dealing with the earth's climate. Each Report was widely circulated for comment to expert reviewers all over the world, usually in several drafts.

This author has been among the expert reviewers for every Report of Working Group I except the first one, and for several of the other reports as well. My comments on the first draft of the most recent Scientific Report amounted to 97 foolscap pages, and related to every Chapter. I claim to have comprehensive knowledge of these Reports, and particularly of their limitations.

The full designation of these Reports is as follows:

- From Working Group I

Houghton, J T, G J Jenkins & J J Ephraums (Editors) 1990 *Climate Change: The IPCC Scientific Assessment.* Cambridge University Press. Referred to as *Climate Change 90*

Houghton, J T, B A Callander & S K Varney (Editors) 1992, *Climate Change 1992: The Supplementary Report to the IPCC Scientific Assessment.* Cambridge University Press. Referred to as *Climate Change 92*

Houghton, J T, L G Meira Filho, J Bruce, Hoesung Lee, B A Callander, E Haites, N Harris & K Maskell (Editors) 1995 *Climate Change 1994: Radiative Forcing of Climate Change and An Evaluation of the IPCC IS92 Emission Scenarios.* Cambridge University Press Referred to as *Climate Change 94*

Houghton, J T, L G Meira Filho, B A Callander, N Harris, A Kattenberg & K Maskell (Editors) 1996, *Climate Change 1995: The Science of Climate Change.* Cambridge University Press. Referred to as *Climate Change 95*

Houghton, J T, Y Ding, D J Griggs, M Noguer, P.J van der Linden, X Dai, K Maskell & C A Johnson (Editors) 2001. *Climate Change 2001: The Scientific Basis.* Cambridge University Press. Referred to as *Climate Change 01*

- From Working Group II

Watson, R. T., M. C. Zinyowera, R.H. Moss & D.J. Dokken (Editors) 1998. *The Regional Impacts of Climate Change.* Cambridge University Press. Referred to as *Regional Impacts 98*

McCarthy, J J, O F Canziani, N A Leary, D J Dokken & K S White (Editors). 2001, *Climate Change 2001: Impacts, Adaptation and Vulnerability.* Cambridge University Press. Referred to as *Climate Change 01: Impacts*

- From Working Group III

Nakicenovic, N. & R Swart (Editors) 2000, *IPCC Special Report: Emissions Scenarios.* Cambridge University Press. Referred to as *IPCC Emissions Scenarios 00*

Metz, B, D Davidson, R Swart & J Pan (Editors), *Climate Change 2001 Mitigation.* Cambridge University Press. Referred to as *Climate Change 01: Mitigation*

"Summaries for Policymakers" and "Technical Summaries" for most of the Reports are available from the IPCC website at http://www.ipcc.ch

A Note on Sources

Climate Change 01 recommends that references should be to each Chapter and they should list all the authors of the Chapter. This has not been done here because of the large amount of space required.

Climate Change 2001: The Scientific Basis has voluminous references many of which have been consulted by this author. The volume itself is an indispensable source of information on current climate science. It is, however, seriously biased in its suppression and distortion of any critical material on the greenhouse effect and its consequences.

Part of this distortion and bias is achieved through the process of *peer review*. The scientists involved with the IPCC have been successful in supplying most of the "peers" that review papers for publication, and in rejecting, or delaying indefinitely, any material critical of the greenhouse effect.

In addition, even when not deliberate, the peer review process can be very slow. Reviewers are not paid, so they are unlikely to rush to review material they disapprove of, and they tend to speed up material that supports their opinions.

Another problem is the escalating cost of official scientific publications. This has reduced access on a colossal scale. Fewer libraries have fewer Journals, and they often restrict access. The cost of interloans has also increased dramatically. Scientists realise this and so try to publish in the more freely available journals Much of my material comes from *Nature* and *Science* which are widely available, and tend to publish any important developments.

The existence of the Internet has only partially alleviated the problem. Very few Journal Editors can afford to publish on the Internet for free. Titles or abstracts may be free, but the full paper requires a subscription. In many cases the old fashioned methods of personal correspondence (made much easier by Email) and requests for reprints, have replaced increasingly difficult library consultations.

The Internet has provided two important facilities without which this book could never have been written.

The first is the free access to recent data. The US Government has been particularly generous in this regard. The following websites provide easy access to contemporary data on the climate, much faster than any Journal.

> The Carbon Dioxide Information and Advisory Center, Oakridge, Tennessee; particularly with their *Trends* facility http://cdiac.esd.ornl.gov/trends
>
> The Goddard Institute of Space Studies, http://www.giss.nasa.gov This site can supply worldwide temperature records in tabular or graphical form.
>
> The National Oceanic and Atmospheric Administration (NOAA) and its National Climatic Data Center (NCDC). Their most useful site is http://www.ncdc.noaa.gov
>
> Carbon Cycle Group http://www.cmdl.noaa.gov/cgg
>
> Satellite Temperature measurements http://www.ghcc.msfc.nasa.gov/temperature
>
> Sea Level Records http://www.pol.ac.uk/psml/programmes
>
> The Climate Research Unit at the University of East Anglia, UK, http://www.cru.uea.ac.uk

Since it is virtually impossible to publish scientific material critical of the greenhouse effect in the established scientific journals, the Internet has provided the opportunity both for publication and scientific discussion which is denied us in the official press. The following websites specialise in such material.

> John Daly "Still Waiting for the Greenhouse' http://www/john-daly.com
>
> CO2 Science. http://www.CO2science.org
>
> The Greening Earth Society http://www.greeningearthsociety.org
>
> Warwick Hughes http://www.webace.com.au/~wsh
>
> Science & Environment Policy Project http://www.sepp.org

A Note on Sources

There is a Yahoo discussion group of climate sceptics with the address climatesceptics@yahoogroups.com

It is unfortunate that the Internet is ephemeral. There is no "archive" of material available, and sites are forever deleting material and changing their address. If the above addresses have changed, try a search engine.

A very valuable source of statistical information is Bjorn Lomborg *The Skeptical Environmentalist* 2001, Cambridge University Press.